OPEN SCIENCE

OPEN | SCIENCE

SHARING KNOWLEDGE IN THE GLOBAL CENTURY

JULIAN CRIBB AND
TJEMPAKA SARI

CSIRO
PUBLISHING

National Library of Australia Cataloguing-in-Publication entry

Cribb, Julian

Open science : sharing knowledge in the global century/
Julian Cribb, Tjempaka Sari.

9780643097636 (pbk.)

Includes index.

Communication in science.

Sari, Tjempaka.

501.4

Published by

CSIRO PUBLISHING
150 Oxford Street (PO Box 1139)
Collingwood VIC 3066
Australia

Telephone: +61 3 9662 7666
Local call: 1300 788 000 (Australia only)
Fax: +61 3 9662 7555
Email: publishing.sales@csiro.au
Web site: www.publish.csiro.au

Front cover image by iStockphoto

Set in 10.5/13 Adobe Minion and Optima

Edited by Peter Storer Editorial Services
Cover and text design by James Kelly
Typeset by Desktop Concepts Pty Ltd, Melbourne
Index by Indexicana
Printed in China by 1010 Printing International Ltd

CSIRO PUBLISHING publishes and distributes scientific, technical and health science books, magazines and journals from Australia to a worldwide audience and conducts these activities autonomously from the research activities of the Commonwealth Scientific and Industrial Research Organisation (CSIRO).

CONTENTS

ACKNOWLEDGEMENTS

The authors wish to acknowledge the invaluable support, encouragement and assistance of the following in this book:

Commonwealth Scientific and Industrial Research Organisation (CSIRO)

National Science Institute of Indonesia (LIPI)

The Rockefeller Foundation Study Center, Bellagio, Italy

Centre for Public Awareness of Science (CPAS), Australian National University

Professor Sue Stocklmayer

Professor Chris Bryant

Dr Per Pinstrup-Andersen, International Food Policy Research Institute (IFPRI)

Dr Bob Clements, Australian Centre for International Agricultural Research

Keith Bashford

Michael Brown

Heather Briggs

Margaret Bryant

Professor Derek Tribe

Nick Alexander

John Manger

Dr Geoff Garrett

Dr Ted Cain

Marie Keir

Wendy Parsons

Jennifer North

Rosie Schmedding

Nick Fisher

Nick Goldie

Dr Tom Spurling

Brad Collis

Peter Cribb

Chapter 1

The case for open science

Nam et ipsa scientia potestas est (Knowledge is Power)

Francis Bacon, *c.* 1620

The need to share human knowledge has never been more urgent. As the world grapples with the acute challenges of resource scarcity, climate change, poverty, ill-health, pollution, rapid urbanisation and food insecurity, it has never needed its science and technology more. However, if anything is to secure the future of civilisation and human wellbeing, it will not be science alone, but the knowledge it yields being shared and employed both widely and wisely. For science and technology to deliver full value to society, they must be accessible to as many people as possible and their messages must be easily understood.

Scientific knowledge is now said to double about every 5 years, but its distribution among the seven billion citizens of Planet Earth proceeds far less rapidly. While the number of scientific papers published grows dramatically with each passing year, the rate at which their essential knowledge is transmitted to ordinary people who might use it in their lives lags far behind. Indeed, it has been claimed that up to half the world's published scientific papers are never read by anyone other than their authors, editors and reviewers – and 90 per cent are never cited.[1]

A vast gap has opened between the creation and the sharing of knowledge. Because of this, a significant part of the world scientific effort is effectively stillborn, or fails to achieve its potential. The intellectual effort, time, money and human genius that is invested in research is lost because of a failure to effectively transmit the fruits of science to the people and places where it is most needed. Scientific knowledge, with the capacity to benefit billions, improve sustainability and protect environments, is often buried in specialised journals, electronic repositories, inaccessible language, IP and legal constraints, or is withheld by privileged elites. Deliberately or unintentionally, barriers have arisen between science and its adoption and use by the people.

This, of course, is not how most scientists would wish it, nor what most governments – who mainly fund the global scientific enterprise – would desire. It is not what industries, which prosper from technical progress, want and it is certainly not the preference of ordinary citizens who depend on science to improve, and even to save, their lives. Yet the gap between knowledge creation and its widespread uptake remains immense.

A reason for this may lie in an observation by the 16th century philosopher Francis Bacon that 'knowledge is power'. During the first 300 years of science, it was generally held that knowledge should be shared freely; bodies such as England's Royal Society and France's Académie des Sciences were set up to foster this ideal: an ideal that remains to this day one of the guiding lights of science – alas, the light is dimming. The 20th century gave birth to the greatest proliferation of knowledge in the million-year story of humanity. Yet, departing from the ideals of the earlier centuries of scientific inquiry, the main driving force of 20th century discovery and innovation was not the quest for enlightenment: it was war. The motor car, the aircraft, the computer, advanced communications, rocketry, modern chemistry, and even aspects of medicine and biology were all widely developed and adopted in service to the military machine. This militarisation of research has largely defined the structure of the modern scientific enterprise. Knowledge, once regarded as the common heritage of humanity, has become the closely guarded asset of the few – a handful of nations, a few corporations, the military and a few elites. More than half a century ago, people in science were already profoundly disturbed by this trend. Sir Henry Dale, president of Britain's Royal Society, said in 1946:

> *I hold it to be our right and our duty to unite in telling the world insistently that if national policies fail to free science in peace from the secrecy it accepted as a necessity of war, they will poison its very spirit …* [2]

The founder of Australia's CSIRO, Sir David Rivett, too, spoke of:

> *… the threat, now much more than a mere threat, to that free trade in scientific knowledge of all kinds, which has been the glory of these last*

> *three hundred years that have seen the most rapid advance in human knowledge of Nature since man began his course.*[3]

While it extended lifetimes and brought great wealth and privilege for one in every ten people, the greatest burgeoning of knowledge has failed, on the whole, to deliver anything approximating a fair sharing of the benefits. One explanation for this is that the system that engendered modern science was shaped, not for sharing and equity, but for exclusion and domination. It is the pressing task of those loyal to the original ideal of universally shared knowledge to lead the change in international attitudes to science and why we do it.

As humanity progresses through the 21st century – the global century – many scholars point to the emergence of a disturbing trend: the world is dividing into those with ready access to knowledge and its fruits, and those without. The people without access to knowledge are not merely deprived of its benefits, they may actually be outcast: playing the role of spectators in the human race rather than runners in it. A former Canadian Government Minister, Pierre Pettigrew, put it this way:

> *In the new economy, the victims are not only exploited, they're excluded. You may be in a situation where you are not needed to create wealth. This phenomenon of exclusion is far more radical than the phenomenon of exploitation.*[4]

The situation is exacerbated by the universal penetration of the media. Satellite TV, the internet, radio, movies, advertising, magazines and newspapers are widening the gap between those with access to knowledge and the power it brings – and those without – in an insidious fashion. Today, the lifestyles of the affluent and their conspicuous and unsustainable consumption are flaunted before all humanity, in virtually every community and in most homes. A consequence of globalisation in communication is the envy and wrath it is kindling between the knowledge haves and have-nots, as well as the colossal over-use and waste of the Earth's scarce resources it is driving. If all humans were to enjoy a lifestyle like that of America or Australia, the Global Footprint Network estimates that it would require the productive capacity of more than *four* Planet Earths to sustain.[5]

This situation presents grave risks for global, as well as societal, stability. Many of the conflicts of recent decades had their deep roots in the scarcity of basic resources, such as food, land and water, as well as political, religious and ethnic disputes that furnish the superficial reasons for conflict. This is due not merely to growing populations and their increased demands, but also to a lack of knowledge about how to use and share resources efficiently, sustainably and equitably.[6]

The nature of 21st century conflict is emerging as quite different from that of 20th century strife, being spurred on by this deficiency in basic human needs, resources and knowledge. Many countries, several regions and some continents now exist in a state of precarious instability as vast pressures build up beneath the surface of societies. Indeed, conflicts have already broken out between the haves and have-nots, the knowledge-empowered and the knowledge-deprived. They are being fought out not on battlefields but in the streets and alleys, the festering shanty towns and struggling villages, the spreading global cancer of drugs taken to blot out the *ennui* of exclusion. In the developing world, this failure to share knowledge fairly causes governments to fail, infant democracies to founder, and unleashes floods of refugees internally and across borders. It was an ingredient in the circumstances that led to the Global War on Terror and instability in central and southern Asia. In both developed and developing worlds, it is turning sections of great cities into combat zones where the affluent inhabit electronic fortresses and the poor and knowledge-deprived stalk streets where police fear to tread.[7]

While one in six humans lives in abject poverty, half the world's people live in a state of knowledge deprivation, meaning that they cannot obtain the basic knowledge or technologies necessary for a decent life, to raise their children, eat well, enjoy good health and improve their circumstances. They also lack the empowerment that goes with solving their own problems.

The knowledge-deprived of the 21st century find themselves at the margins of society: a place where even survival is doubtful for many. The 25 000 children who die daily from malnutrition-related disease[8] also die from a lack of knowledge. The knowledge to save almost all of them exists, but, for various reasons, it does not get through – at least in forms their communities can access, afford or use.

Every year, millions perish for lack of access to affordable drugs for malaria, HIV, diarrhoea, lung infections, flu and other common diseases. This toll has prompted many to question the morality of a global innovation system that sequesters its best knowledge for the rich and powerful.

The knowledge-deprived live in both the developed and developing worlds. They live among us, every day, in each society, almost in every street, suburb or rural village. They include our blood relatives as well as people we have never met. Their only offence is to live 'outside' the great axes of high technology advancement, learning and commerce that now radiate like a giant neural network across the globe. A 21st century world of infotechnology, biotechnology, nanotechnology, genomics, proteomics and bioinformatics is marching on, leaving the majority of people behind.

Science for all of its history has subscribed strongly to the ideal of serving humanity. Nowadays it often finds itself perplexed at the criticism it encounters. Of this, Professor Juan Roederer has written:

> *One would think that scientists have a lot of friends and enjoy public respect. After all, statistics clearly demonstrate that over 50 per cent of the economic growth in advanced countries is based on the application of government-sponsored research.*
>
> *So why is it that in many countries – and most notably the advanced countries – we scientists have no defined constituency, being viewed by politicians as naive, socially ineffective and self important. Why is it that pseudo-science, anti-intellectualism, irrational beliefs and cults flourish like never before? Why is it that universities ... are coming under malicious, sometimes even vicious public scrutiny?*[9]

Roederer concluded there was 'an alarming erosion of public trust' in science, causing many societies and their political leaders to question the motives of the research community, and to impose measures to scrutinise it and even to limit its freedoms. Such limitations, when placed upon science, usually have harmful consequences for free thought, the exploration of ideas, the development of new technologies and the ability of science to make new discoveries.

The 'crisis of trust' in modern science was first brought to public attention by the UK House of Lords in its Third Report on Science and Technology, which recorded 'much interest but little trust' among the British public in science today:

> *Society's relationship with science is in a critical phase. Science today is exciting and full of opportunities. Yet public confidence in scientific advice to Government has been rocked by BSE; and many people are uneasy about the rapid advance of areas such as biotechnology and IT – even though, for everyday purposes, they take science and technology for granted. This crisis of confidence is of great importance both to ... society and ... science.*[10]

Many societies and groups are starting to protest their exclusion from the scientific process. While grateful for the life-saving and life-enhancing benefits of science, in Western democracies the public is already pulling on the reins, resisting the relentless onward thrust of the scientific–industrial machinery, demanding greater scrutiny or placing obstacles in the path of new technologies. Most societies are today questioning the morality, ethics, practices, motives, ownership and control of modern science.

Many reasons have been offered to explain this: the impact of global media and international corporations, the mistrust of governments, professions, institutions and power elites, and the rapid transmission of resistance around the globe. Nevertheless, by an irony of history, the medieval world has somehow been reborn, with the creators and possessors of knowledge and power sequestered behind high walls, and the ruck of humanity outside – excluded yet profoundly affected by what is decided within those walls. In this model, ordinary people are required to pay for science through their taxes, but are often denied information about it, or have little say over its application and control – and are then expected to accept its findings and products gratefully and without question. And they are starting to resent it.

In the early 21st century, there has been a subtle shift in the attitude of society to science in both developed and developing countries. This has moved from a general public acceptance of the authority of science to a questioning of its ethics and trustworthiness. During the Cold War era, science was often identified with national security: it was unpopular, unpatriotic and even personally risky to question it. Science and its secrecy went broadly unchallenged. With the ending of the Cold War, however, science became less closely identified with national security and increasingly aligned with the interests of global corporations, which were the world's new technology powerhouses and research-funding sources. This led to questions in many local communities about whether science was acting in the people's interests – or those of global wealth and power. Individuals willing to tolerate exclusion for national security reasons were not prepared to put up with it for the sake of 'foreign' commercial interests. Coupled with sensational biological experiments, such the cloning of human cells, this has led to the present focus on the morality and control of science.

Early in the 21st century, a European study found a generally positive perception of science and technology (although many citizens regarded it as a sort of Pandora's Box, emitting ills as well as benefits).[11] However, more than 80 per cent of respondents felt that scientists should be compelled by government to respect *moral* standards. The implications are plain: first, that science is no longer generally perceived as an entirely moral pursuit; and, second, if it fails to manage its own ethical standards adequately and transparently, then society will enforce them.

In the developing world there is a parallel situation. Here, Western science is not always widely regarded as being in the interests of the people because it appears, for the most part, as the knowledge system of foreign countries, alien cultures, uncaring corporations or oppressive local elites. Nevertheless, some forms of science involving agriculture, water, public health, transport and the like have been widely applied for public good and have brought very great improvements to people's lives. Thus, it is not the science that is the true object of suspicion, so much as the system that engenders and promotes it.

BALANCING HUMAN DEVELOPMENT

In the morning of the 21st century, knowledge grows faster than anything that humans now produce (with the possible exception of environmental degradation). Faster than food or minerals, faster than manufactured goods, faster than entertainment, faster than money. Since the work of economist and Nobel laureate Paul Rohmer in the 1970s, knowledge has come to be recognised as the primary driver in the creation of the world's prosperity.

With such a surfeit of knowledge, and with such an abyss widening between the possessors and the dispossessed, it is time to contemplate a return to a more traditional ideal: that *knowledge is the common heritage of all peoples.* Not a weapon: a tool of domination or oppression. Not an exclusive possession. Something open to all.

For decades the affluent world has bemoaned the plight of the poor world, yet failed to solve the problem. One reason for this may be the assumption that poverty is a lack of wealth and requires massive transfers of money to remedy it. In reality, poverty more often results from a lack of knowledge. This is the reason it so often appears intractable, despite the millions of dollars thrown at it: money may alleviate the symptoms, but does little to eliminate the causes. Knowledge, on the other hand, empowers people to overcome their own disadvantages and gives them the confidence to do so. Unlike money, it can be shared both easily and freely. The economic miracles of modern China and India both began with the sharing of Western scientific knowledge about food production, and by these countries then applying knowledge in their own ways to create a secure and stable foundation for growth in the rest of their economies. In both cases, agricultural knowledge came first, and was a primary driver in the shift from poverty to prosperity and self-determination.

The successful sharing of knowledge about food production led to the sharing of other kinds of knowledge – health care, industrial and mineral know-how, water and energy, information technology and communications, and a growing awareness of the need to educate all citizens and to protect the environment.

If the world's great challenges in the 21st century are to be successfully addressed, then open science is essential. The cost of this is relatively small and is advantageous to everyone because of the dividends it yields in trade, employment, peace and stability. It has the salient virtue of permitting developing countries to choose those aspects of science and technology that they most need and that best suit their culture, their people, their climate and their environment. If knowledge is widely available within a developing country, it allows individuals and communities to take charge of their own destiny and to build a better future for themselves and their children. This in turn brings prosperity, which can in turn deliver three critical benefits:

- a voluntary reduction in the birth rate, leading ultimately to reduced pressure on key resources such as water and land
- greater political stability and democratisation, resulting in fewer conflicts and refugee crises
- enhanced trade and employment, to the mutual benefit of both developed and developing partners.

The difference between knowledge and money is that money is easily squandered and then cannot readily be renewed. Knowledge, it is true, may be wasted – but once shared, it is usually remains accessible to a community for a very long time and can be applied when required. In the case of the Green Revolution, it is easy to see how the gift of knowledge, adapted for local culture and conditions, can be used by billions of people to build a better future for themselves and their children. It is also clear that knowledge in the hands of billions of people can do more good and generate more economic growth than it can by merely occupying university library shelves or being restricted to a narrow market among the very affluent.

Because knowledge holds the key to wealth and power, as Francis Bacon said, there is a real risk that if the exponential growth of knowledge is confined mainly to wealthy countries, corporations and elites, it will simply widen the gap between the well-off and poor worlds, accelerating the transfer of wealth and resources from the have-nots to the haves.

In its Framework for Action, the 21st UNESCO World Conference on Science acknowledged that, while science and its applications are indispensable for development, the benefits are very unevenly distributed across countries, regions, peoples and the sexes. It also observed that while science has great potential for good, it also has equal scope for harm and so must be embedded in sound ethical principles. It warned that developing countries, especially those rich in biodiversity and natural resources, require special protection from exploitation by wealthy industrial companies from the developed world. It also urged 'better understanding and use of traditional knowledge systems' alongside modern science.[12]

In its closing declaration, the Conference (see Appendix) emphasised four issues:

1. There is a need for a vigorous and informed *democratic* debate on the production and use of scientific knowledge (authors' emphasis).
2. The benefits of science are unevenly distributed; equal access to science is a social and ethical requirement for human development.
3. Science is indispensable to human progress – but its applications can have detrimental consequences for individuals, societies and the environment.
4. All scientists should commit themselves to high ethical standards, based on human rights instruments. Political authorities must respect this.[13]

Only science can deliver humanity from the consequences of the 'big six' crises bearing down on us: the crisis in water; the crisis in resource scarcity; the crisis in land degradation, contamination and species loss; the crisis in food security; the crisis in health; and the crisis in climate change. But none of these can be remedied by governments merely changing a few laws or by companies adopting a few new technologies. Each demands profound change in human behaviour on the part of almost every individual on the planet and, for this to occur, the knowledge of both the problem and what to do about it must first be shared. Science must be open to all.

For example, if climate change could be solved merely by adding geosequestration technology to a few thousand power stations and switching to hydrogen-fuelled cars, it would be fine. But it cannot. It can be addressed only by changing almost every aspect of our lives: from what we eat, to what we wear, how we live, how we raise our children and how many we choose to have, and how we use energy, water and other resources. Such huge behavioural change depends on knowledge sharing on a pan-species scale, rather than on fragmentary technofixes. The same applies to each of the 'big six'.

The problem is that while the world is very well set up to develop scientific solutions and technofixes, it is poorly equipped to open knowledge to humanity *en masse* and universally in forms that they can apply in their daily lives and work. The amount invested in knowledge delivery and adoption is, as a rule, only a tiny fraction of the amount spent on research. Indeed, many of the major problems facing humanity could possibly be solved by applying existing knowledge better and more widely, rather then discovering new – though this should not be taken as an argument to reduce R&D.

The answers to the 'big six' crises now confronting humanity, and which will dominate our destiny in the 21st century, lie not only the creation of new knowledge but more especially in the effective dissemination, sharing and use by people at large of all relevant knowledge. The fate of humanity in this century may well rest on whether or not science becomes more open.

PATENTING AND IP

Patenting and the exclusive ownership of 'intellectual property' is a thorny issue, and it is not the purpose of this book to resolve it. Yet, because this affects the sharing of knowledge in many ways, both positively and negatively, it may be helpful to advance a few principles:

- The private sector and the market are an efficient way of sharing knowledge, and for extending its benefits to the wider community. This will be recognised by any effective science communication and awareness policy.
- Patenting and IP protection are important ways to ensure a fair return to industry for its investment in the research and development of new knowledge and technologies.
- Patenting and IP protection are vital ways to foster continued national innovation.
- IP protection is an important source of revenue for many research institutions, and a stimulus to further research and innovation and to science/industry partnerships.

However, patenting and IP protection has become an expensive industry in its own right, to the point where protecting a technology may cost more than the technology can return. It diverts efforts that should be put into disseminating new knowledge into, often fruitless, legal entanglements. Patents are frequently taken out when a commercially shrewder course would be to be first to market. IP has also become a tradable good in ways that do not reflect the true value of the knowledge to humanity, but rather its value to financial speculators.

Also patenting and IP protection conflict with the principle of the free and rapid sharing of human knowledge. They exclude large portions of humanity from the benefits of science, retard its delivery or price it beyond their reach, they distort the focus of public good research from what benefits society to what is profitable for a few, and they help undermine community trust in science.

In view of these conflicts and contradictions, it seems sensible to seek a middle ground that aims to maximise the benefits to humanity overall. Some ways to achieve this may include:

- restricting IP and patents to novel scientific applications, constructs and technologies
- recognising all chemical elements, genes and naturally occurring materials as the common heritage of humanity, not as private property
- recognising 'primary knowledge' as discovered by basic research as the common heritage of humanity
- building obligations to discuss, inform and educate the community into the granting of IP rights, making the process more transparent and the consequences of new technologies more subject to public scrutiny
- encouraging greater communication by patent holders, using effective communication techniques (such as those outlined in this book)
- encouraging those who take out patents and protect IP to heed the wishes and needs of society, and to engage in an effective two-way dialogue
- developing knowledge-sharing partnerships between science, industry and the community
- creating an international fund to buy out patents, or recompense their owners, in cases where a protected technology is urgently required to save life and deliver large-scale social or environmental benefits in the developing world.

THE GREAT MISMATCH

Universities and scientific institutions worldwide produce an avalanche of remarkable discoveries, insights and advances. However, their capacity to share this knowledge widely with the community, government and industry rarely matches their research skills. Their investment in communicating science is often only a fraction of their investment in discovery. Many invest 100 or even 1000 times more in R&D than they do in transmitting its results and ensuring these are well-adapted to society's needs.

Some people justify this imbalance with the argument that they are research institutions, not communication or technology transfer institutions. In their eyes, their primary role is to discover, rather than to share. Where they do share, it is generally through the scientific literature and their educational activity, although this reaches only a tiny part of the populace. For the most part, scientific institutions are reluctant to invest resources in disseminating the fruits of science, either because they do not know how to or because they regard this as 'a waste of money', or because, bluntly, they cannot be bothered.

Because the public usually funds the science, these excuses are not acceptable. The withholding of knowledge generated with public funds is a form of theft from the people who paid for it, and it is time this moral issue was more widely acknowledged as a prelude to building a more open science.

This book, *Open Science*, is about practical, basic and low-cost ways to share knowledge. It is about developing the awareness of scientific organisations about ways to deliver knowledge more effectively to the society they serve.

Open Science contends that we should be putting as much money, effort and creativity into communicating science as we do into discovery. We should regard those with the skills and abilities to transmit knowledge to where it is most needed as being of equal professional value with those who discover it, as it requires both for science to achieve its full value.

DEMOCRATISATION OF SCIENCE

Earlier we referred to the growing mistrust of science by society, to the increasing significance of ethical issues, to the questioning of the need for change, to the public's fear of alienation and exclusion. These issues can all be addressed by making science a more open and democratic activity.

For the true 'knowledge society' to exist, a cultural change is necessary within science itself, which recognises:

- that the knowledge possessed by the community in the form of values, beliefs, traditions, morality, feelings and behaviours is critical to the successful uptake of scientific knowledge
- that 'lay knowledge' and 'scientific knowledge' are equal, and necessary, partners in the process of innovation and adoption
- that true communication is not just about sharing information, but more about sharing meaning and achieving a common understanding.

Foreshadowing the rise in the democratisation of science, the UNESCO World Conference on Science (see Appendix) said:

> *Today, whilst unprecedented advances in the sciences are foreseen, there is a need for a vigorous and informed democratic debate on the production and use of scientific knowledge. The scientific community and decision-makers should seek the strengthening of public trust and support for science through such a debate.*[14]

The Conference went on to declare:

> *The practice of scientific research and the use of knowledge from that research should always aim at the welfare of humankind, including the reduction of poverty, be respectful of the dignity and rights of human beings, and of the global environment, and take fully into account our responsibility towards present and future generations. There should be a new commitment to these important principles by all parties concerned.*

A conference organised by the British Council concluded that science can, and should, become more open and democratic, and that citizens should be admitted as active partners and participants in the innovation process. It said efforts to promote a democratic science will encourage:

- openness
- transparency
- responsibility and accountability
- independent research and advice
- negotiation of appropriate technological trajectories
- meaningful dialogues
- development of skills and education policy
- forecasting and resolution of conflicts and crises
- equity in the distribution of knowledge and technological solutions.[15]

In *Open Science*, we argue that the democratisation of science is not merely desirable from a societal viewpoint, but also from a scientific one. The community can bring to science many ideas and perspectives that will result in the science being more widely accepted, rapidly adopted or commercialised, and of greater value to more people than would otherwise be the case. Society can be a partner in the process instead of an uninformed, and occasionally reluctant and resentful, recipient.

Democratisation will help to ease public fears about rapid and profound change, and help to allay concerns about loss of control or failure of ethical standards. It will reduce the risk of exclusion. In developing countries it will help bring knowledge to poor people far more quickly by engaging them in the process.

This book is a practical, how-to-do-it guide intended to assist scientific institutions become more effective knowledge sharers and partners, and better practitioners of 'open science'.

We also propose a charter for global science, technology and science communication in the 21st century, appealing to all the world's scientific institutions, scientists, science managers, communicators and policy makers to renew the essential ideal – that science belongs to all humanity – and to join together in bringing it about.

It states:

1. Knowledge is the common heritage of all the world's people.
2. The sharing of knowledge is as important as its discovery.
3. Science will be open. It will engage the community in a democratic dialogue, each recognising the other as an equal partner in human advancement.
4. Partnership between all nations, developed and developing, in knowledge sharing is central to the peace, wellbeing, health and sustainability of humanity.

Chapter 2
Good science writing

Science is by its nature complicated, making it all the more important that good science writing should be simple, clean and clear.

Alas, achieving clarity is something that escapes many scientific writers, whether they are addressing their peers, a knowledgeable but non-scientific audience, or society at large. Indeed, the reader often receives the impression that the writer has not thought much about their audience at all, as they struggled to give birth to long, tortuous and impenetrable prose, with clause piled upon clause, adjective upon adjective, idea upon idea. A good deal of science writing more closely resembles a train wreck than an act of communion with the reader: with words scattered like carriages all over the line.

Good writing begins with the need to pause and reflect on the audience. Who are they? What do they want from your science? How much time do they have for what you are about to tell them? What is their level of literacy or technical understanding? How do they speak and write themselves? What are the issues they are most concerned about or interested in? What will win their hearts or engage their intellects?

Finding out these things requires a skill at which scientists excel, but rarely, in this particular case, undertake: research.

In some situations the answers are easy to come by. Science journalists, for example, usually have a fairly clear idea of their audience, both from surveys carried out by their publisher and from first-hand contact with readers/viewers or receipt of their letters and emails. Technical writers for an industry or professional magazine often have a very clear idea who they are writing for. The communicator for a scientific institution, however, has the challenge of a wide diversity of possible audiences – government, industry, scientists, peers, non-government organisations (NGOs) and special interest groups, the general public and other 'stakeholders' – and has to tune the writing for each audience according to their needs. This often requires research. The same goes for scientists, who are passionate about their work and anxious to share its gems with a wider public: understanding this audience and its needs is an important first step in writing well. For the freelance writer, whose work may end up anywhere from full-length books to short news items, understanding the audience is even more critical, because making a living depends upon it. The advice in this chapter is generic. It is intended for all who write about science, in particular those mentioned above. It refers chiefly to writing for non-scientific audiences – the public, politicians, farmers, industry, and so on – but many of the principles apply equally to good writing in science journals, scientific and institutional reports, and on the internet.

SIMPLICITY IS STRENGTH

Complex ideas do not need to be conveyed through complex writing. Indeed, they are most easily understood by the reader if the language used is simple and clear. This may seem self-evident, but how often this rule is ignored! There are many reasons for this:

(a) The writer fears that simple language will not do justice to a complicated idea (or will 'dumb it down').
(b) Scientific terminology and expression is preferred for reasons of scientific precision, even if it is opaque to the reader.
(c) Writers fail to understand clearly the needs of their audience.
(d) The writer is unaware of how inaccessible professional language is to others.
(e) The writer does not in fact want people to understand what they are talking about because 'knowledge is power'.

The last is a salient and all too common feature of bureaucratic writing, but is also unpleasantly pervasive in the social sciences, education and other specialised research fields, which conceal their meaning behind a vocabulary designed to exclude uninitiated readers. For centuries, lawyers and priests used Latin to invest themselves with artificial mystique and power in the eyes of the public – and some of today's specialists are not much better. However, most of today's science is funded by the public via their taxes – and they deserve an explanation they can understand and, hopefully, make use of.

The true value of science to society depends upon it being explained in a simple, clear way that people can use in their lives, their work or their behaviour. Conversely, science that is explained in an over-complicated or obscure fashion stands a very good chance of never being used, or not being adopted as widely as it deserves.

In short, bad writing wastes good science.

What is bad writing? Well, most people know it when they see it. It is anything that puts a wall of words between you and the meaning. To illustrate the condition, here are two prize-winning examples from *Philosophy and Literature Magazine*'s Awards for Bad Writing:[1]

> *The move from a structuralist account in which capital is understood to structure social relations in relatively homologous ways to a view of hegemony in which power relations are subject to repetition, convergence, and rearticulation brought the question of temporality into the thinking of structure, and marked a shift from a form of Althusserian theory that takes structural totalities as theoretical objects to one in which the insights into the contingent possibility of structure inaugurate a renewed conception of hegemony as bound up with the contingent sites and strategies of the rearticulation of power.*

and

> *Indeed dialectical critical realism may be seen under the aspect of Foucauldian strategic reversal – of the unholy trinity of Parmenidean/Platonic/Aristotelean provenance; of the Cartesian-Lockean-Humean-Kantian paradigm, of foundationalisms (in practice, fideistic foundationalisms) and irrationalisms (in practice, capricious exercises of the will-to-power or some other ideologically and/or psycho-somatically buried source) new and old alike; of the primordial failing of Western philosophy, ontological monovalence, and its close ally, the epistemic fallacy with its ontic dual; of the analytic problematic laid down by Plato, which Hegel served only to replicate in his actualist monovalent analytic reinstatement in transfigurative reconciling dialectical connection, while in his hubristic claims for absolute idealism he inaugurated the Comtean, Kierkegaardian and Nietzschean eclipses of reason, replicating the fundaments of positivism through its transmutation route to the superidealism of a Baudrillard.*

Deliberately, neither of these examples is taken from the scientific literature – to demonstrate to the scientific reader just how impenetrable and exclusive specialised language can appear to the uninitiated. However, there is plenty of science writing that is just as hard for the ordinary person to understand. The flaws are plain: long, tortuous sentences, specialised use of terms and concepts that obscure rather than clarify the meaning, piles of adjectives, pomposity, bombast and a general implied sneer at the intellect of the reader who cannot follow them. This reveals that bad writing can often be offensive as well as annoying. Bad writing is the opposite of communication – which is the sharing of meaning.

In all forms of writing – from poetry to journalism, novels and plays, to speeches and science writing – simplicity is strength. It is the foundation of good communication. Elaboration can come later. A good way to approach science writing is to write the first draft as one would speak to a person very like the audience you are writing for, using their common, everyday language. If this is the general public, write as you would address your aunt or uncle – a person of average intelligence, education and interests, but no scientific background. If the audience comprises politicians or busy senior executives, then write very concisely and with a high degree of impact. These people do not have time to read long, densely argued documents, but generally want only the headline messages. If the audience is farmers, fishermen, miners or foresters, then write as they speak, colloquially and using plain, practical language explaining how the science applies to their activities. Reading aloud what one has just written is a good way of 'hearing' whether it is readily understandable or not. Reading it to a non-scientist or member of the intended audience is better still.

One of the sins of science writing is verbosity – the use of too many words. Because so much science is written in a verbose style, especially in textbooks and journal articles, many scientists find it hard to break the habit. It is, after all, the language to which they are most accustomed; they have had to penetrate its thickets ever since they were undergraduates. Not so the ordinary person who simply finds it incomprehensible and soon loses interest, or is distracted by the effort required to extract the intended meaning from the writing. Scientists sometimes complain that when they explain their work to lay people, they see their audiences' attention wander. The phenomenon is known as EGO (eyes glaze over). It does not mean that the science is intrinsically uninteresting – just that it is being explained in a way that does not engage the audience. This applies to writing, too, but without the warning signs.

The first building block in simple writing is to use short sentences. The full stop is one of the most useful devices in science communication because it allows the reader pause to digest a complex or important idea. This is essential, if science is to achieve full value. Also, it avoids the 'train wreck' of subordinate clauses created by long, turgid sentences, packed with too many ideas and qualifications. It enables the reader to absorb the ideas in bite-size chunks and order them in their mind. It can be used more or less where one would draw breath if speaking the words aloud. It can also be used to create a staccato effect, which is important to high-impact writing, though less desirable in longer articles or documents. The short sentence works well in science journalism, in writing for the internet (where the eye may be tired by sentences that last for several lines), in writing for politicians and executives, and in writing speeches for general audiences. A short sentence usually consists of a subject, a verb and an object. It can have an elegance and an impact all of its own, without having to plunder the thesaurus.

Short sentences impose discipline on the writer. They compel you to ask 'What am I really trying to say here? What is the most important statement to make first, which can then be qualified or explained in subsequent sentences?' Short sentences encourage clarity of thought and expression. In science this is very important because, just as people tend to form judgements about other people if they wear ragged clothes, speak badly or have poor personal hygiene, people also form judgements about science based on how well it is communicated. This is human nature, and there isn't much one can do about it. Clear, concise writing suggests that the scientist has thought clearly about the issue concerned. Turgid, abstruse and laboured language, on the other hand, conveys an impression of muddled thinking and not caring much about the reader: it does the science itself a disservice, which in turn may limit its value (as well as the prospects of its researchers).

The art of writing a short sentence lies in reducing the number of subordinate clauses. This means keeping to a minimum the number of clauses beginning with 'which', 'that', 'who', 'when' and other qualifying words. For example, we could easily have written the previous two sentences as a single sentence of 35 words, with the word 'which' joining them. We chose instead to break it in two, without harming the meaning but slightly improving the clarity. Short sentences do not devalue science. They enhance it. Using short sentences also obliges the writer to decide which is the most important fact and present it first, instead of running everything together in a single sentence and making the reader guess.

As with roses, the secret of concise writing lies in hard pruning. After producing a first draft, it is essential to go back over it and strip out every needless word or phrase. You may be surprised how often three, four or five words can be replaced by a single word. Believe it or not, there are few pieces of scientific writing that cannot be improved by removing half of the words initially used. Try it! When writing about science, prune, prune and prune again. Eliminate all extraneous expressions, clumsy phrases, non-essential adjectives and adverbs, and obscure or bureaucratic terminology. Rephrase more economically. Then, having stripped the writing to its bones, you can return to it to elaborate, as required. This is how to write well: building the edifice on a plain but strong foundation.

The gardener often enjoys the act of pruning, knowing it will result in a good crop of flowers. Likewise, the act of pruning one's writing can be enjoyable: seeking to convey the absolute essence of what one is describing. Many people, including many scientists, find writing an unpleasant chore, rather than the fulfilment of their research. Pruning can make the labour pleasurable, as well as enhancing the meaning and significance of the science itself in the minds of those who read about it.

CURING OLD VICES

Common vices in science writing include the use of the passive voice instead of the active, the use of the subjunctive mood instead of the present or future tense, the over-use of adjectives to describe a single noun, and the use of professional terminology or 'jargon'. It is quite easy to purge oneself of these bad habits without having to go back to school to study grammar and syntax.[2]

A great deal of science is written in the passive voice, rather than the active. The active expresses the action directly: 'We pursued the research'. The passive focuses on the object being acted on: 'The research was pursued by us'. The reason for overusing the passive voice probably lies in the desire of scientists to appear objective and impersonal when describing experiments and their results. However, science uses the passive to gruesome excess; this makes the writing ponderous and less easily digested than it should be. It adds unnecessary words – in the above example, 50 per cent more words are used by the passive. Writing for the public should avoid the passive voice as far as possible (e.g. instead of saying 'The passive voice should be avoided in writing for the public ...'.). Even scientific editors no longer favour the passive. Search for it in your writing and convert it ruthlessly to the active voice. Your prose will sparkle with new vigour and directness.

For example: 'In this study the chemodynamics of heavy metals in soils were investigated.' Why not simply 'In this study we investigated the chemodynamics of heavy metals in soils'? Or instead of 'A new treatment for diabetes has been developed by Australian scientists', just write 'Australian scientists have developed a new treatment for diabetes.'

The use of the subjunctive mood is a common feature of science writing, which makes it more turgid and its meaning more vague and uncertain to the reader. Without getting into technicalities, the subjunctive is characterised by the use of words like 'would', 'could', 'should', 'may' and 'might'. These are often preferred by scientists to the use of the present tense (is, are) or the future tense (will, shall). However, they increase uncertainty in the reader as to what is meant – and removing them often does little damage to the sense. For example, in the sentence 'Heavy metals could pollute soil or groundwater ... ' the word 'could' can be omitted: 'Heavy metals pollute soil or groundwater ... ' This is simply a cleaner, more direct way of writing, which avoids the subjunctive but does not significantly alter the intended meaning. It expresses the meaning more directly and with less uncertainty.

Of course, science often wants to convey a degree of uncertainty, and this is the reason for the ubiquitous 'could' and 'would'. However, this is often faulty reasoning on the part of the writer. Uncertainty can be conveyed directly by stating that the conclusion is not certain, or open to different interpretations, and explaining why. This is more direct and honest than using syntax to obscure the meaning, and the reader will appreciate it. Where it is unavoidable, the word 'may' is often preferable: 'The universe may end, not in a bang but a whimper ... '

STRUCTURING THE ARTICLE

The traditional scientific journal article begins with a few general statements about things that are usually well-known or accepted. It then outlines the background to the research, provides a description of the experiments carried out and their methods, reports and discusses the results, then finally draws a conclusion from them and discusses its wider implications. The reader must work their way through each of these steps in order to be rewarded with the finding.

A science article written for the media or a lay audience, on the other hand, adopts almost exactly the opposite structure. It reports the main finding and its impact on society in the very first sentence, then explains who did the research and why, adds further detail and finally, if there is room, goes on to discuss what most scientists would see as the main game

– the research itself. This is because audiences are usually more concerned about how the science affects them directly than they are with the method by which it was achieved. They are users of science, not its practitioners.

In journalism, the conclusion is nearly always presented first and the rest of the article then expands on this, providing the reader with the supporting evidence for the initial claim and the background to it. This structure has since become common in many forms of reporting: corporate and government reports, for example, present their findings in an executive summary – often a series of terse bullet points – so the busy reader can seize the essence without having to wade through the detail. In journalism, most readers read the first few paragraphs, but few make their way to the end of the article. If important information is placed here it will be lost (or even cut out completely by the editor).[3]

This 'upside down pyramid' article structure, with the most important fact first, achieves a much higher impact on the reader and is likely to stick in their mind longer. Where there are several important findings from the research, the article will present them one at a time in the first few paragraphs, rather than risk obscuring or losing some key points by running them all together.

Scientists often assume the reason they are doing their work is self-evident, but this is often not the case. A good science article therefore makes clear, in its opening paragraphs, *why* the research is being carried out – to save lives, prevent environmental damage, improve industrial productivity, and so on. Indeed, it is on this simple fact that the importance of the article and its chances of publication depend. If it is omitted, the relevance of the science to the reader may well be lost. The editor may regard the story as unimportant and 'put it on the spike' (discard it).

The credibility of science with the public often depends on who performed it, so the science article identifies the researchers and institutions involved early on. This is a sign to the reader – who may be unfamiliar with journals and peer review – how trustworthy the information is. However, a good article or media story does not waste space on long wordy names, titles or teams.

A good science article often goes directly to the meaning of the science to society, rather than to the science itself. This is especially the case with a new technology or piece of applied science. The exception would be a 'blue sky' discovery, or findings from fields such as astronomy or palaeontology without immediate practical application. In these cases, the article will dwell on the sheer wonder or novelty of what has been found and seek to engage the reader through their curiosity about the natural world.

To engage the reader at the outset it is vital to choose a strong heading. Unlike a scientific paper, where the heading often describes the research, a heading in the media, a press release, a book or a report is intended to catch the eye and capture the attention of the reader – not to inform them. It is usually concerned with the impact of the science, not with the science itself. It is an advertisement for what follows, not a synopsis of it. For this reason, a strong heading is usually short – three to five words work best. An attractive heading may also use mystery, humour or an unusual word to attract the reader. All it needs to do is entice them into reading the first paragraph, which then delivers the main message of interest and lures them to read on.

An effective piece of science writing often has only one idea per sentence. As mentioned above, this gives the reader time to digest important facts.

Where high impact is required, occasionally use only a single sentence per paragraph.

The white space between the paragraphs emphasises the point being made in a delicate way, without using exclamation marks, underlining, **bold type** or *italics*. In fact, the last three can offend the reader, as they are the typological equivalent of shouting at them (like using CAPS in an email).

Good science writing is usually very economical in its use of language. It compensates for complexity by elegance and simplicity of expression and choice of words. It avoids pomposity or talking down to the reader. It goes directly to the wider significance of the research and why it was done. It explains its relevance to the general reader, rather than to science. It seeks to convey a sense of wonder, where appropriate, but does not exaggerate or overstate. It is checked with the scientist, to ensure accuracy. If there is room, it refers to doubts, criticisms and alternative interpretations of the science.

JARGON AND BUREAUCRATESE

Scientists use language in very particular ways in order to convey specialised meanings. This works fine among the peer group but it can lead to confusion, ambiguity and misinterpretation externally. Because science itself is concerned with being as precise as possible, it is a great shame if it loses precision because its audience misunderstands what they are told. In science, new words are often coined to describe new phenomena, or else old words are given new meanings to which the public is not privy. Scientists sometimes forget this.

For example, a soil scientist may refer unthinkingly to a soil layer as a 'horizon', whereas his lay audience may wonder what that line the sun goes behind is doing at the bottom of a hole. This is a case of one word having two (or more) meanings: technical and general. Classically, scientists often refer to their 'models', blithely unaware that many people in society think a model is an elegant person sporting stylish clothes, or alternatively, a small plastic aeroplane. In the sentence 'We are using a new model to predict rabbit populations…' the average person may be puzzled why the scientist would employ a mannequin to forecast rabbit plagues – and probably wonder how the scientist came by such a generous budget!

Scientific terms slip off the tongue, or the keyboard, very easily, and great care must be taken to avoid them or at least to translate them for the audience. Is a base an electron pair, a headquarters or the bottom of something? Is a phase part of a waveform cycle or a period in your life? Is a port the plughole in a computer, a place for ships to dock or a fortified wine? Is a bond a chemical link, a financial instrument or a manacle? Context will usually supply the answer, but one can never be sure what all readers will make of it and science writing must always be scanned carefully for such ambiguities.

A good test for whether a word is jargon is to imagine oneself standing at the supermarket checkout and saying the word to each person as they come past the cash register. How many would be able to provide even a rough explanation of the meaning? If the answer is 'not many', then the term should be avoided and a more common term used.

Avoiding scientific jargon is not as hard as it seems, as articles written for the public, for government and even industry usually focus on the application of the science, not on the science itself. It is nearly always possible to describe the application of science in plain language. Nevertheless, scientists sometimes complain that the translation of science into plain language 'devalues' it or 'dumbs it down'. However, if the use of scientific terminology will only cause the audience to misunderstand – or, worse, completely misinterpret what is being said – then it makes no sense to use it, as the result will only be confusion. Scientists should never expect people outside their discipline to understand the exact meaning they ascribe to a specialised term – even an apparently simple one like 'model'. Every effort should be made to re-phrase the language so that it has meaning for the audience. This sometimes takes more time and effort than some researchers can spare, and is the reason for the growing value of the skilled communicator as a messenger and interpreter between science and society.

Another challenge for the science writer turning scientific reports or articles into stuff the public can understand is 'bureaucratese': the leaden language of the public servant. Nowadays science is often twice as difficult to understand because it mingles scientific jargon with bureaucratese. This language is supposed to be dispassionate, but in fact it is usually clumsy, verbose and hard to read. It too favours the passive and the subjunctive, as well as a whole lexicon of specialist terms intended to exclude the uninitiated. Indeed, bureaucratese is often deliberately designed not to be understood, or else to be ambiguous, in order to withhold knowledge (and power) rather than share it. Because a great deal of science happens in bureaucracies – in universities, research agencies or government departments – the two languages often become horribly intermingled, resulting in a disaster for clarity and for the communication of science. In writing about science, it is very important to purge bureaucratic language as well as technical terms.[4]

A nasty bureaucratic habit is to refer to everything by its initials or its acronym. This is fine if you know what it means – but is simply gobbledegook to the general public. Acronyms are bad in several ways: first, because they are meaningless by themselves and cannot even be looked up in a dictionary; second, because the phrase from which the initials are drawn is usually badly chosen and not easy to guess; third, because acronyms break the flow of meaning by forcing the reader to pause and puzzle over them; and fourth, because they sneer at the person who has not been initiated into the secret of their meaning.

A related phenomenon, even where the acronym is explained, is 'alphabet soup' – the excessive use of initials, as in the following example:

> *The FAIMMS sensor network will utilise leading edge technology to provide real-time 3D profiles of reef systems at seven sites along the Great Barrier Reef (GBR). AIMS is the national operator of FAIMMS, which is one of the components of the Great Barrier Reef Ocean Observing System (GBROOS), for which AIMS is also responsible. GBROOS is part of a nation-wide collaborative program, the Integrated Marine Observing System (IMOS), designed to observe the oceans around Australia.*

It is possible for the general reader to fathom what is meant here, but the over-reliance on obscure abbreviations creates constant hiccups in the flow of meaning and should be avoided.

Another common vice of scientific (and bureaucratic) writing is to attach too many adjectives to a single noun. Sometimes as many as five, and even seven, adjectives may be piled onto one poor, struggling, inoffensive little noun. The words 'one', 'poor', 'struggling', 'inoffensive' and 'little' are the adjectives that describe the word 'noun'. The use of such strings can perplex the reader, who has to decide which adjective is the most important in the context, and how each adjective affects all the others. The use of too many adjectives to over-describe an object is bad writing and unnecessary. If the adjectives are essential they can be distributed over several sentences. In reality, however, most of them can be left out without losing meaning. This improves both clarity and ease of reading. When pruning one's work, it is good practice to remove all adjectives. Then go back and see which ones are truly vital and allow these alone to stand.

WHO? WHAT? WHEN? WHERE? HOW? WHY?

A good science article answers all these questions. Especially it answers the question 'why?', explaining to the reader the reason the research was carried out, and why it is important to humanity. The reason why is often taken for granted in scientific writing, which is a big mistake when writing for a wider audience.

Each question helps to establish the meaning of the science to the reader. 'Who', for example, explains who is affected by the science, and who performed it. This conveys both its relevance to society or industry, and its trustworthiness, embodied in the name of the research institute or corporation. 'What' explains what was actually done. 'When' conveys to the reader whether this is new knowledge, or 'news'. 'Where' is important because people habitually think of their own locality first, and science performed locally by local researchers addressing local problems is of much greater interest than science performed in some other country by and for people they have never heard of. 'How' explains how the science was actually performed and how it affects the community.

LIVELY WRITING

Good science writing contains passion. In this respect it is quite unlike scientific writing, where the goal is to be objective and engage the reader's mind through fact alone. Science writing seeks to engage both intellect and feelings, making it one of the higher literary forms. It should intrigue and inspire, and provoke surprise, wonderment, fear or excitement. It can be clean, elegant, even beautiful. It can have rhythm and music.

Passion can be displayed in many ways – in the choice of words, the vigour of the prose, the cadence of the sentences, the use of metaphor and analogy, and the colour and pace of the language. Just as we are engaged by a lively speaker more than by a dull one, science can hold greater significance for the reader if the writer allows their feelings to show. This is why it is often a challenge for the trained scientist to make the conversion from the cold, dispassionate prose in which findings are habitually reported to the livelier style required to convey it to a wide audience. However, many accomplish it with triumphant success – Jared Diamond, Richard Dawkins, Stephen Hawking, Stephen Jay Gould, Paul Davies and Tim Flannery, to name but a few.

Books, magazines, newspapers, the internet, TV and radio are nowadays chiefly designed to entertain – and when science appears in them, it too must entertain as well as inform. It should court controversy rather than avoid it, as debate is the fuel of democracy. It should present itself in lively ways, with plenty of direct quotations (as distinct from indirect quotes or citations), because the spoken voice lends vividness and immediacy to the subject. It should employ familiar imagery from sport, the arts or daily life to help make the unfamiliar familiar.

Good science writing uses punctuation thoughtfully, to aid the reader's understanding by pausing in the right places and avoiding long, complex sentences. Commas and full stops both achieve this and can be used plentifully (though not before conjunctions such as 'and' and 'but', which are supposed to join sentences together). There is a lot of confusion over the use of colons and semicolons. As a rule, a semicolon can be used to divide a sentence more strongly; this avoids breaking it in two; all parts of the sentence should have a verb. A colon can be used to highlight what follows: as illustrated in this sentence. It can also be used at the start of a list of facts or statements.

Exclamation marks rarely have a place in science writing as they give it an exaggerated, rather shocking quality, which is not always in keeping with the serious nature of science and responsible writing. Used sparingly, however, they can arrest the reader! They can also convey humour, irony or strong emotion.

Do rhetorical questions have a place in science writing? We would say that, on the whole, they do not, for the simple reason that they look as if the science writer is asking the non-scientific reader a scientific question. This is both silly and irritating. Rhetorical questions are fine in rhetoric, where the speaker may use them to keep the audience awake. In prose they are a distraction. They are more a feature of the undergraduate essay than the well-constructed science article or news report. Science is all about asking questions and seeking answers to them, but these questions do not have to

TIPS FOR GOOD SCIENCE WRITING:

- In science writing, meaning is more important than style.
- Choose language your audience uses habitually. Relate the science to their interests, not yours.
- Simplify first, ornament later.
- Prune up to half the words you first wrote. Strip out adjectives and adverbs. Replace only those most essential to the meaning.
- Rewriting is as important as writing. Work through the first draft quickly and then come back to tighten and improve it.
- If you are struggling with a sentence that has become too long, kill it and start again. Waste no time on a sentence that has gone wrong.
- Write several short sentences in preference to one long, cumbersome one.
- Don't start important writing late at night or when tired. Sleep on it and allow your subconscious to order and structure it.
- The hardest paragraph is the first one, as it has to catch the reader's attention. Don't let it put you off. Write whatever seems good, then press on with the rest of the article and return to rewrite the first paragraph. Rewrite it four or five times until it says exactly what will capture the reader's attention.
- Avoid the passive. Avoid the subjunctive. Avoid too many adjectives and adverbs.
- Use short sentences and short paragraphs. Punctuate to give time for thought.

be rhetorical – that is, addressed directly to the reader. They can simply be framed as the question that the scientist is trying to answer.

Another common error in science writing is the use of 'statements of the bleedin' obvious'. Poorly written science articles often begin with one of these (perhaps because scientific papers usually do, as they proceed from what is well known to what has just been discovered). In science writing there is no such need. In fact, the use of such statements interferes with the telling of the story as they force the reader over dull and familiar ground, rather than telling them at once what is new and fresh. In a world awash with new information, concealing the main point of a science story behind obvious or well-known facts only reduces the number of people who will read it and use the knowledge. Most do not want to waste time reading about things they already know, so they rapidly browse on to something more stimulating. Statements of the bleedin' obvious are the enemy of good science writing and should be avoided, especially in journalism and media releases.

WRITING FOR THE INTERNET

Writing for the internet is not so different to writing for the print media. Prose that is short, crisp and clear works best on a computer screen. Concise paragraphs, plenty of headings and white space avoid tiring the reader's eyes and hold their attention far better than slabs of type.

Internet users are quite often young and impatient, rather than old and persistent. They scan, rather than read. This means that the writing should be seeded with words that will catch their attention and invite a closer look. Today's internet user is like a grasshopper with a short attention span, skipping from item to item or site to site, often in microseconds. Dull writing and text-heavy layouts cause them to skip away. As a rule, good web writing:

- is short, sharp and snappy
- is fresh and newsy
- has the main point at the top
- uses dot points and crisp delivery of facts
- has informative headings
- has an open, attractive layout
- contains reputable information sources
- links to places the reader can find out more
- is more concise than scientific and print media writing
- has one idea per paragraph
- is illustrated with attractive thumbnail images.

Much is written about website design, but as a rule a clean, uncluttered layout with plenty of white space works better for science than one jam-packed with small items, boxes, ads and gizmos. It looks more authoritative and less amateurish. It conveys information more efficiently. It is less difficult to maintain and keep in good order.

The essence of the internet is that it is a public highway. One of the worst crimes habitually performed on it is to create a website that invites the public in – and then closes the door in their face. This poor outcome is achieved by:

- making information hard to access and contacts hard to find
- hiding phone numbers and email addresses
- concealing the identity of staff, providing poor searchability; making the site look stale by failing to refresh its content
- using clunky online forms
- having poor navigation and searchability

- putting firewalls and password protection everywhere
- using poor, uninformative writing
- generally making visitors feel they are unwelcome.

With those caveats, the internet is one of the most important outlets for good science writing and open science – and one that by 2010 reached over two billion people all around the world.

Chapter 3

Planning for open science

Science communication is all the better for a well thought-out plan. Science is a very focused business – so focused in fact that the views, needs, wishes and values of the wider society are sometime downplayed or set to one side. Yet very little science and technology is widely adopted without the sanction and active compliance of society, or at least of government and industry. Indeed, in modern democracies citizens frequently exercise a right to veto or regulate any science or technology they have doubts about. Scientific and technological progress is therefore not the outcome of vigorous pushing by scientists, as it is sometimes depicted – but rather of an active dialogue and consensus between science and society, in which each understands the other. The society that does this best advances the fastest.

Having a sound communication plan helps scientific managers, scientists, innovators and others to see their discoveries or technologies from the perspective of society or users, rather than of science alone, which is a very healthy thing. The reason is that scientists, technologists and engineers are often passionate about their work, which tends to make them less than wholly objective when it comes to discussing its pros and cons with the wider community, who may have doubts or questions about it. Left unresolved, these doubts can grow into active public opposition, resulting in regulatory and even legislative controls, moratoria and even outright bans on certain research or technologies. Genetically modified (GM) food, nuclear energy, stem cell science, food irradiation, xenotransplantation and more recently coal-fired energy and certain nanotechnologies are all examples of science and technology to which many societies, and indeed whole countries, have objected in various ways – despite the obvious benefits of some of these things. This phenomenon is known as technology rejection, and it is increasingly common as democracy expands, as the rate of technological advancement increases and as science communication itself falters.

Society knows from long experience that most technologies turn out to have downsides. Indeed, much of the global scientific enterprise today is occupied with cleaning up or rectifying the results of the use, misuse or overuse of previous science and technologies. The public, in other words, expects most new science and technology to go wrong at some point, often at the cost of human life and suffering. Science, on the other hand, prefers to look on the optimistic side and seldom makes allowance for the public's natural caution, apprehension and wisdom when promoting its latest idea, discovery or advance. A good science communication plan can, however, go a long way to improving both society's understanding of the new science – and science's understanding of the wishes and attitudes of society. Between the two, this smoothes the path for the adoption of the new science or technology and reduces the scope for unanticipated consequences and harmful side effects.

The word 'plan' is used here in preference to the more commonly employed term 'strategy', which comes from the Greek 'strategos', meaning a general. Strategy describes the general's art, which is to kill the enemy and win the war. It is not a very appropriate term for science communication because it embodies ideas of violence, predomination and conquest. Good communication is always a sharing of ideas and meaning, in which messages, opinions and information from all sides are received, considered and discussed until a common understanding of what they mean is attained and a way forward found.

It is usually the goal of a science communication plan to bring together the public interest with the scientific potential. However, the public may hold a very different view of the science to that held by its proponents, and this needs to be borne in mind when designing communication. *Science communication is not propaganda for science*, yet sometimes scientific organisations suffer a regrettable tendency to thrust their opinions on society without bothering to find out or consider how society will react to them. This may create obstacles to the ultimate adoption of the science, as it is seen to suit the needs of scientists or corporations, of elites, rather than those of society. On the other hand, the most effective science communication is usually that which is closely tailored to society's wishes and needs, and is well-informed about these.

Good communication also needs a unique feature to attract attention. The unique qualities possessed by a scientific institution or a corporation are its research discoveries, technologies, achievements and, in some cases, individual research staff. Others may work in the same field but nobody else can precisely emulate these. So, the core of good a science awareness plan turns on this principle: the public reputation of a scientific institution and its staff rests on the effective communication of its real achievements, their meaning and benefits in a dialogue with society. A bad plan brags, boasts and makes hyberbolic claims about the institution and its science and generally ends up making both look dodgy.

Conventional use of commercial branding techniques to present an artificial image is generally not very effective in science and may even prove harmful to the research institution's public reputation, as it treats scientific claims the same as commercial claims and opens them to exactly the same suspicion in the public mind. It is also very expensive and, although a corporation may be happy to spend millions on burnishing its image, few science agencies can afford to do so.

Hence it is both wiser and more economical to communicate the science rather than to promote the institution. If the science is good and delivers clear benefits, the institution's reputation (or brand) will prosper accordingly. It will not gain from telling a sceptical public how great it is – a lamentable error that contaminates much of the communication of universities and other scientific institutions nowadays. Boastfulness is not attractive in an individual, let alone an institution. In science, scientific impact is the brand – not the big PR claims. Science is trusted by the public because its findings are independently and objectively reviewed by scientific peers; the claims of commerce are rarely reviewed, tested or validated in this way and often prove wanting. It is sensible, when communicating science, not to muddle the two.

The chief value of a good science communication plan is that it helps the institution and its science managers to examine their science from the standpoints of their audiences – the public, industry and government, or subsets of these. It encourages them to find out what potential users and beneficiaries think of a new science or technology, whether they want it or not, how they would like it delivered and what their reservations or questions about it are. So, the first task in designing a communication plan is to identify who are the audiences and what are their needs. Unfortunately, this is performed ritually in many cases, without careful inquiry into what particular audiences actually think, need and expect of their science provider. The people inside the ivy-clad walls sometimes assume – on the basis of their specialist expertise and personal experience – that they are also conversant with what people outside the walls feel, think and say. The irony of this is that science – the great gatherer, analyst and user of data – often proceeds to deliver its products without collecting or analysing any data whatsoever about how the public or particular groups of customers will receive it. There are gaping holes in science's understanding of society, and filling them is seldom catered for in research budgets when, in fact, it should be one of the first questions asked before research (especially applied research and publicly funded research) is undertaken. For economy's sake, science often compromises by relying on second-hand information about public attitudes taken from industry or government, for example. But industry consists of companies, many of which eventually go broke because they fail to understand their own markets – so this source of information is not always reliable. And governments are run by politicians, who often see only what they wish to see and ignore that which they do not – and invariably end up losing office because they are out of step with the public. Neither source can be completely trusted. A good science communication plan seeks to close the gap in understanding of what society wants from science by doing the necessary research. Techniques for doing this are discussed in Chapter 4.

Having carried out some research into public (or industry) attitudes to the proposed science, a useful way to orientate the communication plan is to identify the main national benefits (social, economic or environmental) that it will deliver, and to test these with audience samples to see that they are credible. Apart from the fact that triple-bottom-line goals (economic, social and environmental benefits) are much approved by the wider community, this approach allows the research body to demonstrate its commitment to the national interest or public good – even if the immediate beneficiary of the research outcome happens to be a company, an industry or a government agency. Thinking about national benefits and national priorities applies a sensible discipline to the framing not only of communication strategy but also research strategy. It keeps the focus on the ultimate value and impact of the delivered knowledge rather than its technical aspects or organisational preferences.

Communication plans should always be prepared by people who know something about communication, rather than by an internal committee. Internal committees often consist of people who know little about communication, though they may be well-qualified in other fields. Nonetheless, they will *all* have strong views about communication: it's one of those skills that everyone with a degree reckons they possess, but all objective evidence points to the contrary. Internal committees are also driven by internal agendas and how they wish to present their organisation or own role to the outside world. In other words, they don't actually want to *communicate* so much as foster a cherished illusion about themselves; that is, commit propaganda.

Scientists are among humanity's most passionate enthusiasts. They adore their occupation in a way that surpasses most other workers. When you speak to them about it, after being assured of your genuine interest, even the driest and most pedantic researcher undergoes a startling metamorphosis into an evangelist. You can't blame them for wanting a say in how their work is communicated. They are seized with a vision of having other mortals experience the joy, exhilaration and enlightenment that understanding a tiny piece of our wondrous universe confers. And they often confuse the wish to share their love of their work and their desire to promote their institution with delivering the science in ways that society can understand and use. For these reasons, it is wise to keep the main communication planning in the hands of those who have either done it before, or have at least some practical experience such as regular contact with media, industry or government.

Military strategy and communication planning are alike in that no plan survives first contact with the real world. The need for flexibility, fluidity and constant intelligence of what is going on are paramount. A detailed, step-by-step, carefully timetabled plan is a straitjacket. It will have you making your grand scientific announcement on the day the general election is called, a new tax is announced or a war breaks out. It will advertise your lack of connection to the immediate concerns of society and the 'real world'. It will reinforce the widely held impression that scientific institutions are medieval monasteries that are shielded from the outside world and isolated from its urgent cares. This will weaken the inclination of taxpayers, governments and companies to fund your research. More importantly, it will undermine your efforts to communicate because society will be wary of institutions that appear out of touch.

There is, of course, a place for timetables in a communication strategy, especially if you are pursuing an approach whereby successive announcements build on one another, or if you have a deadline to meet. However, these can easily be derailed by external happenings. It may be more effective to try to time your announcements for moments when the public, industry or political debate is focused on a related issue or perhaps when it is subdued. That is when media, government, industry and the public will be most receptive to what science has to tell them. You should avoid if possible having your communication plan driven by internal needs (such as achieving goals by a set date so that a report can be made) at the expense of effective communication that is sensitised to the needs of the audience and is released at the right moment.

DESIGNING A PLAN

Designing a communication plan is not a black art. It need not cost a fortune nor take months. It should not be long and tedious. The best ones are concise, compelling and dynamic: a few pages at most. They are simple, logical and easy-to-follow. They enable basic communication activities to cascade through the organisation, involving staff in straightforward activities that are effective. They teach, lead and guide, rather than direct and enforce. Above all, they encourage the organisation to gather external feedback, develop the skills to interpret it and respond intelligently.

Here are the basics. The exact order may vary according to the situation.

1. **Define your overall communication goal(s)**. Make sure they are practical, achievable, relevant to society and can be evaluated in some way. Carry out the research necessary to find out how these mesh with the goals of society or your target audiences.
2. **Identify target audiences**. Find out what it is they want or expect from science. Find out what they know and don't know of your work. Find out how they would prefer information to be delivered so they can use it. Find out about any concerns they may have that might derail the delivery of the scientific outcomes.
3. **Segment audiences** into key groups, each requiring different communication methods and messages.
4. **Work out the relationships** you wish to have with each audience. Relate them directly to your institution's overall goals or business plan. Aim to build trust, mutual understanding and a two-way information flow.
5. **Decide key messages**. These depend on the audience's outlook and will vary from one audience to another. The messages should be few, simple, and should either be overt or implicit in all communication activities you undertake with that audience.
6. **Choose your tactics**. Based on your research with each target audience about its needs and preferences, work out the most effective ways to reach them and to engage them. Make sure communication runs in both directions wherever possible, and that there is provision for feedback.

7. **Identify the resources** (money, staff, external skills and equipment) you need for each tactic, and budget for them. Timetable where appropriate.
8. **Develop ways to evaluate** how your program is going, to correct anything that seems not to be working and build on tactics that prove successful.

Table 3.1 offers one possible structure for an organisational communication plan. The upper part is the bit that receives management sign-off and should remain consistent over time. The actual tactics (or initiatives) are the part requiring flexibility and operational latitude to cope with shifts in audience opinion and external circumstances.

MANAGEMENT AND COMMUNICATION

The goals in a communication plan very often spring directly from the goals stated in the institution's strategic or business plan – its purpose, mission and specific objectives. They go at the top because the communication plan is derived from them.

However – and this is a serious issue – organisational purpose, mission, vision and goals should not be developed without careful thought given to how they will appear to, and be received by, the outside world, and without some market testing.

Institutions sometimes fall into the trap of setting themselves missions and goals that sound fine internally but look appallingly self-interested and out of touch when viewed externally. In other words, communication principles (what do our publics and stakeholders expect of us and how do they value us?) should guide the formation of institutional objectives, rather than internal opinion. Public perception needs to be integrated into the organisation's overall planning and management from the outset if the aim is to see its science and technology more quickly and widely adopted. A communication plan alone cannot correct institutional goals that are poorly chosen or out of step with public values or expectations.

Communication, both internal and external, is intrinsic to the image of the organisation and cannot simply be bolted on after management has decided where it wants to go. Communication consultant Jim Macnamara describes the persistent failure of managements to grasp the function of communication as 'wheel-trim' syndrome: you can add drag-stripes, fancy paintwork and trim to your car but they won't make it go any faster. Only if improvements are built into the actual engineering of the vehicle will its

Table 3.1 Shaping a communication plan

Communication plan
Purpose A concise statement of the purpose of the institution – its vision or mission statement, expressed in terms relevant to external audiences and stakeholders
Strategic objectives The institution's main strategic or business goals
Communication objectives Goals for the organisation's communication activities. These underpin and support the strategic goals, above. They reflect the needs of customers, audiences, partners and stakeholders, as well as the organisation itself
Communication strategy A short statement of the main methods and principles by which the above goals will be achieved
Principles Principles that all communication activities will observe and uphold
Key elements Describes briefly how the organisation will go about implementing its communication plan
Ethos Core beliefs of the organisation it wishes to project externally
Target audience *Government* politicians, senior public servants and relevant agencies *Industry* industry groups, professional bodies, regions, research partners, etc. *Internal* staff, scientists and other professionals *Public* general public, consumers, gender, age group, educational level, income group, etc. *Research partners* other universities, research agencies, government agencies, commercial partners, scientific bodies
Key issues Issues emerging from dialogue or market research with each segment that affect the relationship, research adoption and communication methods
Key message(s) Messages based on audience research and the viewpoints of research users
Initiatives A list of methods designed to communicate with the target audiences: • face to face • publications • electronic • media activities • advocacy

Communication plan
Measurement Measures by which the success or failure of individual initiatives can be assessed: • industry survey • economic impact analysis • staff survey
Budget Funding allocated to each activity in this part of the communication plan • revenue
Timeline Sets deadlines and dates by which certain tasks are to be accomplished
Staff and resources Staff and resources, including external, allocated to this part of the plan

performance be enhanced; in the same way, only if communication is built into management, will management be effective.[1]

International mining industry public affairs authority George Littlewood adds:

> *Too often the business of managing public perception, and therefore the public clout of the organisation, is in reactive mode, driven by day-to-day events, not long-term planning. It is far more important to do the opposite, to be proactive. The key is to listen to what's going on external to the organisation.*
>
> *It is about engaging both critics and supporters and, importantly, it is about the public affairs function being the sophisticated and penetrating eyes of the organisation. What is going on in the wider world must be better understood within the organisation so that its response and its management can be better tuned to external pressures.*[2]

COMMUNICATION OBJECTIVES

The goal of most scientific organisations is to see their work adopted or commercialised for the betterment of society. The communication plan should reflect this by bringing the views of society into the process of delivering relevant, high-impact science and technology.

The fundamental rule in setting communication goals is to reflect not only the interests of the organisation, but especially the interests of the customers, stakeholders and the public. Plans that place the interests of the institution ahead of these three are liable to make the organisation appear selfish and remote. Business talks about creating 'shareholder value'; science should speak in terms of societal value, whether it is discussing commercial research or research for the public good.

However, science often serves two distinct stakeholder interests simultaneously – the interest of the immediate customer (perhaps a company, a government or a funding source) and the interests of the wider community. Sometimes these interests conflict and the research institution is compelled to ask itself where its true loyalty and obligations lie. In a world of tight funding for science, the overwhelming temptation is to follow the dollar but, as has been mentioned, this is leading to a crisis of trust for science and its reputation for caring for the public good. In the longer term this may erode both public and private support and retard the adoption of beneficial new research.

It is not as hard to reconcile the interests of commercial customers with those of the wider public as might be imagined – and industry will often benefit from it. This issue is dealt with more fully in Chapter 6.

COMMUNICATION PLAN OVERVIEW

This part of the plan states the main methods and principles by which the communication objectives will be achieved. It consists of a few concise but clear statements of purpose, or broad objectives, for the communication strategy.

For example, in the case of a scientific organisation, such purposes might be:

- to help reduce degenerative disease by providing better information about nutrition and exercise
- to enhance economic growth by developing and communicating more efficient production and processing techniques
- to introduce industry to new environmentally sustainable technologies that are cost-effective
- to influence public behaviour towards a more conservative use of water and energy
- to influence government and industry to adopt a more scientifically sound greenhouse policy.

Having selected a few key communication intentions, it is then important to state the principles, key elements and ethos on which they are based. The reason for this is that all human decisions are based on a combination of rationality and emotion: it is essential to acknowledge the emotional basis of one's intentions as well as the objective basis, as it provides the mainspring of motivation. In an organisational communication strategy they may comprise:

- a vision statement (what, idealistically, we hope to achieve)
- a statement of principles
- a statement of ethos or core beliefs of the organisation that we intend to display through communication
- a statement of key elements (broadly, how we intend to go about it).

TARGET AUDIENCES

The next step is to identify target audiences, markets or 'publics'. In communication, there is no single 'public'. There are scores, possibly even thousands, of different publics, all of whom react differently to information presented to them and who have special needs and interests of their own, often differing widely from one another. A classic case is the contrasting attitudes of industry, government and environmentalists to, say, greenhouse issues. Although the science may be the same for all, the way it is interpreted and communicated can be quite different for each audience.

In a communication plan for a scientific institution, typical broad audience categories are:

- government
- industry
- the general public
- external research partners
- staff and key stakeholders
- non-government organisations.

However, within each of these categories there will be quite distinct audience groups. The public, for instance, may consist of consumers, householders, males or females, the young or elderly, urban or rural dwellers, professionals or unemployed, and so on. Each group may require its own opinion research, messages and communication techniques.

The next step is to work out the relationship you want to have with your particular target audiences. Do you wish to learn what they think? Do you want to engage them in discussion, inform and raise awareness, increase the uptake and application of a technology, or generate more investment in research? Once again, the relationship should be defined in terms of the audience's needs, as well as those of the research institution.

The critical thing is to research the audience to find out their needs and expectations from your work, what they know and don't know, how they prefer to receive information in a form they can use, and so on. Different techniques are discussed in the Chapter 4. Failure to carry out this research at the outset will limit the success of the communication strategy and the rate of adoption of the science. It is like trying to perform scientific research without data or measurements.

KEY MESSAGES

Key messages are the things you want to say constantly and repetitively to your audience – in different ways and either directly or implicitly – in every communication you have with them.

Science prizes originality and novelty and shuns constant repetition (which it often regards as tantamount to plagiarism). In communication, repetition is essential if you want people to remember what you tell them. You say the same thing as many times, to as many audiences, in as many different ways as you can imagine – and eventually it may become public opinion. (This may take time. In climate change science, the production of new evidence starting in the mid-1970s, and constant repetition, only began to achieve traction with governments and the public in the early 21st century – just over a generation.)

Key messages come first from what your audience wants (or needs) to hear – not from what you most want to say. *Then* they come from the overall communication objectives. If you neglect the audience's needs, you will impair your whole communication effort. It may leave the scientific organisation looking propagandist and self-centred.

For example, in speaking with consumers you discover they have a great anxiety about the need to improve food safety and purity. In speaking with your staff you discover they have a great anxiety to be seen to be doing brilliant food research. If only the second point emerges in the publicity it is likely to miss the mark with consumers, who generally don't care about research and have no way to judge its brilliance. However, by structuring the communication strategy so that the key message speaks of discoveries and advances that significantly improve the health and safety of food for consumers, both goals are achieved. The public will be pleased their food is safer, and the researchers will be recognised as having made a vital contribution. This may seem blindingly obvious, but it's amazing how often it is overlooked.

The same applies equally in communicating with industry or government. Research among politicians may reveal their prime concern is to show their electors that they are doing a good job in order to secure re-election. Scientists, as before, want to show that they are doing great work that deserves more funding. The solution may be to try to identify the political benefits in the scientific success – how many jobs it created, how many lives were saved or improved, and what the gains were to the electorate. These are the sorts of things that politicians delight in taking ownership of, and in sharing the credit with the scientists who actually made the discovery! As a bonus, the politician then becomes a *de facto* science communicator for your team.

A company facing tough local or foreign competition may be primarily concerned about its ability to compete. What it is looking for from the scientific institution is that its researchers have the ability and track record to give the company an edge in the marketplace – not that they are producers of elegant science.

These examples underscore the principle that *the key message must be drawn from the perspective of the research user, not the research producer.*

There may be one key message or several. All should be simple, clear, non-technical and customer-focused. Too many messages may confuse the audience. These messages will be delivered in different ways in all communication with the target audience: be it verbal, printed, implicit in media coverage, encased in submissions, reports, brochures and marketing material, on websites, and even on institutional gifts and greetings cards.

Feedback from the audience should be encouraged so that the key message can be continually fine-tuned in keeping with shifts in external opinion and customer demand. Changes in customer opinion and the key message must also flow back up the research chain so that they can be factored into the R&D itself and help to iron out bumps and potholes in the pathway to adoption.

COMMUNICATION INITIATIVES

These are the heart of the communication strategy and cover the full range of activities designed to convey information or impressions to the target audience and to gather feedback.

These initiatives may employ the familiar tools of glossy publications, mail-outs and advertising; they may cover face-to-face and interpersonal approaches; they may use the internet including social networking sites and multimedia; or they may use indirect means of reaching the audience such as media coverage, advocates and ambassadors. They will be dealt with more fully in later chapters, but some useful general points follow.

Publications

Publications consist broadly of reports, discussion papers, brochures, handbooks, directories, annual reports, leaflets, capability statements, fact sheets and calendars. They are a good way of presenting the organisation's identity, capabilities and achievements, but they are greatly over-used. Their first drawback is that they rely on the recipient having the time and interest to read them carefully, which is seldom the case in a world awash with information. Their second drawback is that they are, too often, designed to gratify the internal corporate ego rather than to inform the external recipient. Their third drawback is that, in presenting too slick a corporate image of the institution, they may in fact undermine trust, credibility or its image as a cost-effective operator. Private companies and research funders are allergic to their money being spent on glitz; environmentalists and taxpayers hate waste paper.

Scientific organisations suffer from publication mania. Perhaps this is because the scientific paper gives substance to the work of research, so the corporate handout is seen as somehow giving substance and credibility to the institution. Generally, when faced with a communication task, the Pavlovian reaction of scientific managers is to demand a brochure. These are churned out with a freedom, a frequency, at a cost and with a wastefulness that would be considered shocking in a private company. Science, it may also be argued, is among the world's biggest environmental vandals when it comes to felling forests to satisfy its craving for corporate gloss.

To avoid this, every proposed publication should be subject to a strict justification – not just churned out at the whim of a manager. The rules for a brochure costing thousands of dollars are no different, in essence, to the rules for launching a national newspaper costing tens of millions of dollars:

- Who is it aimed at?
- Do they really want it?
- Will they read it?
- What is the competition for reader time?
- Is it the cheapest way to deliver the organisation's message?
- Does it fit the communication plan?
- What are the cost/benefits of doing it?
- Do we really need it?
- How will we know, objectively, whether or not it is delivering our science to users in a useful way?

Scientific organisations, which are usually so careful with public money, experience a rush of blood when it comes to corporate publications. They churn them out willy-nilly, on a management impulse, without discipline or analysis, at a cost of tens of thousands of dollars each. One of the communicator's hardest tasks is pouring cold water on corporate enthusiasm when in full cry for yet another pointless but ego-gratifying publication.

Face-to-face communication

Face-to-face communication is almost always the best and most effective way to get a message across, listen to the audience and build the relationship. It may consist of one-on-one meetings, larger gatherings such as advisory bodies, seminars, workshops, exhibitions and presentations, all the way to major conferences. Its salient advantage is that it allows knowledge and opinion to flow both ways.

For face-to-face communication to be truly effective, however, the scientific institution needs to have highly developed listening skills, negotiation skills and presentation skills. Some scientists have told us that among the most satisfying meetings they have ever attended were ones in which the scientists never spoke a word, but simply received the views of their research partners and external people and took note of them. This 'biting the tongue' technique is not only educational for the recipients but it also transmits an enormously positive message about their openness to others' views, needs, criticisms and preferences. Helping one's interlocutor to feel 'heard' is one of the most important skills in modern science management and communication. Even more importantly, it values the knowledge that exists in industry, government or the community, and places it on an equal footing with the knowledge of the scientific body. This is an absolutely critical step in reaching a common understanding, enabling knowledge sharing to speed and smooth the adoption of scientific findings.

One good, but rarely used, opportunity for the science community to communicate face to face with external stakeholders is the futures conference. Practically everyone has an interest in the future and some sort of view of it, whether they regard it as a threat or a promise. Much of the public's interest in science lies not – as scientists sometimes believe – with an intrinsic fascination with the research, but more in a kind of horrified apprehension about what is coming next: What new toxin will appear in our food? What disease or pollution scare will surface? How will new technologies change our lives for the worse as well as the better? How will it harm the environment? How might our jobs, communities and values be threatened by technological change?

The futures conference explores scenarios, both good and bad, that science predicts on the basis of its latest advances, discoveries and insights, and combines them with what industry and the community imagine and wish their futures to be like. Its value is to help people to prepare themselves for the future by sharing their impressions of it, and begin the process of adopting scientific findings by giving thought to what is coming their way. The value to researchers is that these conferences can flag unanticipated social, ethical and other concerns about the science, in time to do something about them, as well as making them more conscious of public needs and wishes. They are also valuable for identifying novel applications of science not yet contemplated by researchers.

If a university or large science agency were to hold futures workshops across the full breadth of its disciplines, it could develop an interesting blueprint for national or regional development – a road map to the future.

Electronic technologies

Electronic technologies offer an avenue for providing, acquiring and exchanging knowledge on an undreamed-of scale. They include television, video, radio, the internet, email, mobile communications, electronic conferencing, virtual workrooms, multimedia, games, haptic devices, holography, and the like. In recent years, the growth of social networking on the internet has opened up a whole new realm for the potential communication and discussion of science.

Electronic technologies have the virtue of being able to reach thousands, or even millions, of people, but they have the disadvantage that many of these technologies are entirely optional on the part of the recipient. They are not, for example, all that precise a way of communicating with decision makers and opinion leaders, who are usually too busy to surf the web, check email or watch hours of television – although these technologies are starting to have an appreciable impact on politics via the general public. Despite various advances in the interface, they are also impersonal, lack human warmth and interaction and are open to abuse. They still pose problems with the key goal of communication: achieving shared meaning. Many technologies are dominated by giant corporations whose motives and intentions are suspect to the community. Some also have the drawback of being alienating to certain classes of user, such as the elderly, technologically phobic, and the vision- or hearing-impaired. Also, they do not reach the poorest and most needy members of the human race, who have most to gain from having greater access to knowledge.

One of the most attractive features of electronic communication is its cheapness compared with face-to-face techniques or publications. However, it is also quite hard to measure the impact of broad-scale electronic communication, and therefore to establish its true cost–benefit relative to other methods. Web 'hits' do not measure the information absorbed and subsequently used by the reader, and neither do TV ratings. On the other hand, email contact and virtual conferencing can be very precise and generate useful feedback. A sound course is to employ electronic communication as part of an overall spread of techniques for reaching particular audiences.

Advocacy

Advocacy is when another individual or organisation carries your message to the target audience. This can apply to a story run in the media, a satisfied customer speaking to their peers, a politician sharing credit for your work, or a famous or influential person recruited to promote your work or raise public awareness of it – such as Al Gore in the case of climate change.

As detergent companies have known for generations, it is also highly effective, because your message comes with someone else's endorsement and avoids the appearance that you are merely blowing your own horn. One of the mysteries of the modern media is that while people affect to despise it, they also have a tendency to credit what they read, hear or see in it, and the fact that your story appeared in the paper is likely to do more for your credibility than telling it yourself.

To make best use of it, advocacy involves having both the most credible message and messenger. However, both must be carefully chosen to suit the audience at whom they are aimed. A rock star may have more credibility with a youth audience than a Nobel laureate!

MEASURES OF SUCCESS

Measures of success are an important way to test the effectiveness of your tactics, and of your overall strategy. They are also important in demonstrating to the institution that its investment in public awareness is paying off. Ideally, they should be expressed in quantitative, as well as qualitative, data. This is because *scientists trust numbers more than they do words or images.*

Measures of success can range from evidence of greater awareness among certain audiences, to customer satisfaction ratings, increased adoption of advice or technology, and greater public and political consensus on a way forward.

Many people use a 'value of publicity' assessment, which attributes a dollar advertising value to the volume of media coverage achieved. Although the numbers can be impressive, this is generally not the most appropriate tool for expressing the effectiveness of a science communication activity. After all, what does the fact that you gained $1 million worth of free publicity actually tell you about the uptake or use of knowledge?

Measures of success are *not:*

- 'we produced a brochure'
- 'we issued six media releases'
- 'we gave five presentations to industry'
- 'we created a website'.

These are merely measures of communication output, not of success. They don't tell you a thing about what the outside world made of you and your work.

PROGRAM AND PROJECT COMMUNICATION STRATEGY

An essential element of an organisational communication strategy is that its principles, objectives, messages and tactics cascade downwards within the organisation like a fractal series. The communication strategy for a large scientific program, or for a small scientific project, should be miniaturised versions of the overall plan that have been fine-tuned for particular audience needs.

Negotiating this across a large organisation can be one of the trickiest tasks because some scientific cultures (e.g. sociologists, agricultural scientists, astronomers and environmental researchers) are habitually communicative, while others (e.g. earth scientists, mathematicians, physicists and engineers) tend to introversion and prefer to communicate among themselves. Some disciplines have a high public-interest focus, while others observe a culture of commercial secrecy (sometimes, alas, employed as an excuse not to communicate). The best move is to design several simplified copies of the plan according to the different needs, skills and communicativeness of the various departments, units or teams.

The principle here is: *to achieve optimal awareness of its value to society, a scientific institution should seek recognition for all of its work, not simply for a part of it.*

Most institutions are more renowned for one field of activity than for another. They may have a high profile for their medical research but a low profile for their metallurgical work, even though the latter may be just as excellent. Good strategy will aim to preserve the high profile for medicine while enhancing the profile of metals research.

Metals may, at first glance, seem a lot less sexy than developing a new cancer therapy, but they too save and enhance lives, create jobs and income, and give rise to new tools and technologies and things the community can relate to. They also often mean more to industry. All that is required is a bit of imagination in how the work is explained, and a rigorous focus on application and its consequences.

Developing a communication plan for scientific programs and individual projects follows the same basic steps as for the organisation, although it may be far less elaborate and use fewer tactics:

- define the communication goal(s)
- identify target audiences and their needs
- decide key messages
- choose your tactics and timetable
- identify the resources and budget
- devise a way to measure results.

In a scientific organisation, it is highly desirable for every program to be required to communicate, and for each scientific team to go through the process of thinking about how it will deliver its findings or achievement to various audiences. This helps the organisation to focus on customer and societal needs, as well as its own, and on scientific impact as well as research. This aids the ultimate goal of successfully transferring knowledge and technology.

It ought to be mandatory for universities and other institutions operating under a competitive grants regime to include a basic communication plan in every grant application, and for its implementation to be a condition for obtaining future grants. In block-funded institutions it may be necessary for management to send clear signals throughout the organisation that teams and individuals are expected to communicate. Ideally, this is coupled with a system of incentives, rewards and promotion. All science has an obligation, at some point, to 'report to the shareholders', and it is beneficial to foster a communicative culture. Scientific bodies wishing to maintain or increase public funding of their work need to be especially mindful of this.

A critical element in developing awareness plans for individual research programs and projects is to have communication expertise available to guide the process. Although many institutions employ a science communicator, they are normally few when compared with the number of projects to be communicated or the number of scientists requiring advice and assistance. This is one reason why it is highly desirable to equip scientists with the skills needed to communicate their own work more effectively to different target audiences, such as media, government, industry and the general public. Part of developing an awareness plan for a scientific project consists of identifying which members of the team will make good spokespeople, and then ensuring they have the necessary training to do the job with confidence and skill.

Universities, unfortunately, are generally poorly equipped when it comes to science communication skills. There is usually a public affairs office, whose task is mainly to market the institution and its educational function, but not all employ a trained science communicator, let alone several. The result is a marketing office that has few of the necessary skills and, in some cases, may not even appreciate the difference between science communication and institutional PR.

One reason universities struggle for funding is their lack of skill when it comes to showing their real value to the wider community, government and industry. Many university systems, somewhat complacently, assume that their contribution to society is self-evident and that only a fool wouldn't see it. Alas, this is not always the case. A second reason is that, in marketing themselves, they may be seen to serve their own needs rather than the needs of the wider community, and so miss the mark.

A lack of funding and support for universities and research systems stems primarily from society not being aware of the genuine contribution they make – and so is unable to exert pressure on politicians. Politicians, who like to spend money on things that make them popular and get them re-elected, don't feel compelled to favour universities and research agencies to win voter support. However, if the university or research agency can, with evidence, convince the public of the lives it saves through its science, the jobs it creates, the environment it enhances, the wonderful new products it designs, it will stand a far better chance of winning public and political support than the timeworn (and often unexplained) assertion that we need to give more money to science or to universities.

Chapter 4
Understanding the audience

In communicating science, nothing is more important than the ability to understand one's public, to listen to their views and values, to obtain feedback and to measure the effectiveness of one's communication activities. These are areas of research and measurement that commercial and political organisations perform as a matter of course, and think nothing of spending millions of dollars on them. They appreciate their importance. Science, the great researcher and measurer, largely does not.

As a rule, scientific organisations rarely devote more than a fraction of their resources to this vital activity. However, if the goal is to share knowledge more efficiently and widely, to assist the smooth commercialisation or adoption of research findings, to build external trust and increase investment in science, this is a strategic error.

In the 21st century the old 'we know what's best for you' model of knowledge delivery is no longer acceptable. In educated democracies it is becoming less and less so. Democratic societies have their own views about what is good for them and have little compunction in rejecting science, technology or anything else that does not meet their standards or their needs.

All good communication plans contain a research component: finding out first about the audience and what they want, and second whether or not the communication is working. Because funds for this sort of research are invariably tight, the purpose of this chapter is to explore low-cost but effective means to obtain feedback from various audiences, measure their responses and better understand external perceptions of your institution and its work.

Just as most of us would have trouble in recognising ourselves from a rear view, many scientific bodies find their external image quite unfamiliar and strange to begin with. It can be an educational experience finding out what a largely uninformed society actually thinks they do behind the ivy-clad walls or cyclone fences, and what it makes of their work.

Nothing beats face-to-face feedback and exchange of information. However, you can only do this with a limited number of people; it takes time, costs a lot and can produce distorted results. The big risk is that science managers seeking feedback may come back from a foray into the outside world laden with what they wanted to hear. Feedback obtained in this way should always be compared with an objective source, such as data from opinion research with various key audiences.

One of the most cost-effective ways to obtain feedback on an institution and its work is what we term 'trilateral' research. This presents three separate, overlapping perspectives of the organisation. It consists of:

- quantitative opinion research across large samples of the population and specific target groups
- qualitative (or focus group) research using small but representative groups to help interpret the numerical data obtained from the first method and to identify perceptions, ethics, values and drivers
- media analysis, which analyses the image of the organisation and its work as presented to the public and key target audiences through mass, specialist and local media.

Both quantitative and qualitative research can be carried out among the general populace or among specific groups such as customers, partners, decision makers and opinion leaders. To identify issues of concern and needs, it is highly desirable to generate both macro and micro views of the organisation. For example, consumers may tell you things that fill them with concern about a new technology – which the industry hoping to implement it may fail to mention (as was largely the case with GM food).

Scientific organisations generally have little difficulty with the notion that industry has specialist knowledge that needs to be exchanged with researchers in order to produce a satisfactory partnership and research outcome. However, they often doubt that the community at large has special knowledge that is essential to the science process. Yet it does. This 'knowledge' is embodied in community views, morality, needs, perceptions, traditions, fears and concerns – all highly unscientific things on the surface (though they have kept the human line extant for several million years). Nevertheless, for a new scientific advance or technology to be adopted requires both good science and societal acceptance. Indeed, society can often provide sage advice about what makes a technology acceptable or unacceptable.

The failure of science to consult society is one reason for the growing crisis of public trust in science described in many recent scholarly studies,[1] and for the paradox that the increasing commercialisation of science appears to be accompanied by decreasing public confidence. Ways to address this are discussed in subsequent chapters, but broadly they involve science communicating not only with its immediate partners in industry or government, but also with the wider community – as it is they, in the end, who will decree what industry and government decide to do.

Scientific organisations need to be highly attuned to both their immediate customers and the society they ultimately serve: they need to 'listen with both ears', not merely with one. They need to accept that useful knowledge and expertise reside in the wider community, as well as in industry and science. They need to view the public as partners in the innovation and knowledge-sharing process, not as a surly Luddite mob that enjoys trashing exciting new technologies.

Furthermore, if science truly understands what consumers or the public want and believe, then it can greatly help industry to achieve a commercially successful research outcome (or government to achieve a better policy) and so become a truly constructive partner. On occasion, science will also need the fiscal guts to tell industry that some technological advance it wants is likely to prove unacceptable from a social perspective – and suggest a better way to do it.

QUANTITATIVE RESEARCH

Quantitative research is also known as opinion polling or public opinion research. It is normally carried out by professional pollsters and market research companies, sometimes by the sociology and politics departments of universities, and occasionally by government agencies. It consists of a series of simple questions – usually with yes/no or good/bad answers – that yield a numerical result such as '75 per cent of the population has heard of your institution (men 78 per cent, women 73 per cent), but only 15 per cent know of your work in food research or marine biology'.

The advantage of this technique is that, for a comparatively moderate cost, it can give you a snapshot of what the population, or subsets of it, knows and does not know about your organisation and its work, and how highly they value it. The most efficient way to use this technique is to employ a professional pollster and to tack your questions on to the end of the regular questions they ask of the population, such as 'Who do you intend to vote for?' and 'What make of car do you plan to buy next?' The cost of doing this varies according to the number of questions you pose and their complexity, but is usually some tens of thousands of dollars and more if you commission exclusive research. It has the advantage of using a well-characterised sample of the general population – as small as a thousand individuals – that is continually being checked and refined for its representativeness.

The focus of the polling can be on the institution itself, but should preferably be focused on the issues its research seeks to address and to gauge the level of public interest in them, and in the proposed solutions, to assess the degree of public support or possible opposition. Opinions derived from such polling can be analysed according to gender, age cohort, geographic locality, income group, and so on.

As a one-off, or 'snapshot', this sort of information may not appear very useful except to let you know where you stand at a particular time. Repeated at regular intervals, however, it soon gives a clear idea of whether public awareness of your work is growing or declining, whether approval is rising or falling, where you need to concentrate greater effort and more resources, and what issues may be brewing – in short, the basic information needed to finetune your communication plan and make your science as acceptable as possible to society.

As a cross-check, it is advisable to include the same questions in polling of important groups – clients of the organisation in various industries and sectors, politicians, media editors, public opinion leaders, senior bureaucrats and the like. These groups both lead and follow general public opinion closely, and any differences can be meaningful. If the public, for example, strongly supports research to overcome a particular problem about which politicians are indifferent, it can be used to sharpen political awareness of the importance of the science: politicians hate to feel out of touch with community views on anything. A contemporary example of this was the use of public opinion polling on climate change, which has persuaded more than one reluctant government that if the public were worried, they had had better be worried also.

Regular quantitative analysis also has the virtue of providing a basic statistical underpinning for the communication plan, an objective way of measuring its effectiveness and progress.

QUALITATIVE RESEARCH

Qualitative (or focus group) research is one of the most valued techniques in the marketing industry for understanding how consumers may react to a new product, the positioning of a brand, and so on. Scientific organisations tend to dislike it because its data are 'soft' and non-numerical. Yet it reveals an extraordinary range of emotional, psychological, intellectual, educational, sociological and cultural issues that go into forming public opinion, trust and the decision to accept or reject a particular technology or piece of scientific advice.

One of the most useful experiences the management of a scientific institution can have is simply to sit and listen to a group of 'ordinary citizens'– unbriefed and not particularly scientifically literate – discussing the institution and its work. It can open their eyes to social and political realities in a way that mere statistics cannot.

A focus group is a discussion among a representative group of citizens on a topic of interest to the body that commissions it. Although the discussion may be facilitated, a fruitful approach is not to pose questions but rather to let conversation flow among the group on the general topic, allowing them to show what they do and do not know and how they feel about an issue. The results can be presented in a statistical way, but as the sample size is small – generally a dozen or so people in a group, with a limited number of groups – it is wise not to place too much reliance on this. A cross-check with your numerical survey will tell you whether the focus group accurately reflects wider opinion.

However, what you are really after are comments, remarks, criticisms, quotes, ideas, (mis)conceptions and moral and values statements – the things that tell you what the people are really thinking. George Littlewood recounts how the mining industry for a long time misled itself by relying only on statistics that said that the public rated its economic contribution above its social or environmental responsibilities. It wasn't until they did focus research that they found, with shock, that the public actually expected it to deliver all three. Focus work lets you see and understand things that aren't obvious in the raw figures.

Another example is five successive opinion polls taken in Australia over a decade that show the public to be more interested in science stories in the media than they are in sport, politics, economics, crime or employment. On the surface this might be taken as very encouraging sign for science. However, when the qualitative work is done, it emerges that many people are actually frightened of science and the change it brings to their lives; they are nervous of technology rather than excited by it. Thus their high 'interest' in science may be more of an apprehension than an enthusiasm. If you never did the qualitative work, you might assume from the opinion polls alone that society was strongly pro-science – and then you would be shocked when it rejected the latest technologies. Using focus research to probe beneath the surface of public opinions allows you to explain what drives them – and this is generally a complex and ever-changing mix of positive and negative perceptions.

Another use for focus research is in the naming of scientific institutions and products. What sounds a perfectly sensible professional title within the discipline may send all sorts of confusing or alarming signals to the community, and you'd never know it until it was too late. Worst of all, it may send signals that you are arrogant, remote, aloof, disinterested in community values, tuned-out or irrelevant – and not worth investing in. What ordinary citizens think may seem of small importance but, sooner or later, it is reflected in what politicians think too. Choose your name with care – and check with the public.

MEDIA ANALYSIS

Media analysis is an increasingly refined and useful tool for monitoring a scientific institution's overall image, as presented to society or segments of the community through the media. It is particularly useful for tracking issues over time and judging how the climate of popular opinion is evolving. Its greatest value, however, lies in the ability it confers to finetune external awareness strategy and tactics, right down to the level of local media and individual journalists. It is a first-rate way of staying flexible and in touch.

To appreciate the value of media analysis requires the institution to have a reasonably sophisticated understanding of the media – to grasp that what appears in it is not there merely because of journalistic whim but because society applies market pressure to the media to run certain sorts of stories. The media, in other words, are a mirror of society: often close to the leading edge of shifts in public opinion but also rarely actually inspiring them because it is unwise in a business sense for the media to get too far ahead of their market. The reason for science to analyse its presence in the media is to help it understand not only how it is being presented to the community, but also where the leading edge of public debate is heading so that research can keep in touch and stay relevant.

Media analysis not only quantifies the organisation's presence in the media (number of stories, length, financial value and source) but also assesses their quality. This is a complex art, for which there are various different formulas, but basically it involves a combined assessment of:

- the quality of the medium in which the story appeared and its audience reach

- the position of the story within the medium (e.g. an early newspaper page or the 'back of the book'; the 6 o'clock news or a midnight chat show)
- the nature of the overt and implicit messages that appeared in the story (did it convey the desired message to the audience?)
- factors that influenced readership or audience reach (e.g. position on the page or in the program, accompanying visual material, reputation of the journalist, editorial treatment and headline).

The data also reveal: who are your most active and effective spokespeople and critics; which media give you the most prominent or sympathetic coverage and which were most negative to what you are saying; which journalists are most supportive or critical; which issues are bringing credit to you, and which disrepute; and what are the messages most commonly presented about your work through the media.

Media analysis helps you to understand the origin of shifts in public opinion discernible in opinion polling or focus groups. For example, a science agency knew from opinion research it had high awareness in one city and declining awareness in another. It turned out that the leading daily newspaper in the city with poor awareness had a very poor coverage of science. This enabled the communication plan to refocus on the particular paper to redress the problem.

In an even more cogent example, successive polls over a number of years revealed declining science awareness and interest among people aged 14–24 years. Media analysis showed that most science coverage was in print media, but that young people simply weren't reading newspapers like their parents had done. Instead they were getting their science from TV and the internet. This prompted a major shift in communication tactics to target electronic media.

Media analysis is invaluable in managing a 'crisis'. Not only does an institution confronting a crisis need to know *what* is being said in the media on a daily basis, it also needs a more objective measure of how the issue is developing than one can get just by scanning headlines. Media analysis provides an index of how favourable or unfavourable media coverage is, both for the organisation as a whole and for a particular issue. It will reveal what are the positive or negative messages about your organisation emerging from the coverage; it will also give you a better idea of how prominently the issue is being presented, which helps you to react appropriately (and avoid over-reaction). It tells you which journalists and media you need to build better relationships with. Long term, it can tell you when an unfavourable issue has finally slipped out of public consciousness or been replaced by a more positive portrayal of your science.

TRIANGULATING THE IMAGE OF SCIENCE

The use of qualitative, quantitative and media analysis together offers the most cost-effective way for a scientific organisation to judge its own reputation, and the value and acceptability of its science to the community or industry it serves.

Together, these three methods enable the user to achieve a 360-degree view of what people think about it, or an issue, in numerical terms, what underlies their opinions and how the world at large perceives it. It allows the user to measure these views as they change, and to understand the reasons for the change.

Above all, it allows the scientific organisation to couple the science it delivers more tightly with the wishes and needs of society – and so have a far greater and earlier impact. It enables science to deliver greater value to society.

CUSTOMER VALUE ANALYSIS

Customer value analysis (CVA) is one of the more business-orientated methods for assessing the value that proximate (i.e. immediate) customers see in and obtain from a scientific organisation. It can be applied to virtually any client group – industrialists, service providers, government, farmers, health-care workers, environmentalists, and so on.

It may employ a spectrum of both quantitative and qualitative tools, from questionnaires to face-to-face interviews, web surveys and focus groups, and range from medium cost and broad spectrum to expensive and highly targeted. Its greatest strength is the focus on client needs, especially the form in which they want new knowledge and technology to be delivered. Its primary concern is with producing as seamless a 'fit' as possible between science provider and science user.

Questions posed to the customer typically ask them to rate, and comment on, the price and quality of features of the research provider and the partnership such as:

- level of expertise
- application of relevant skills
- quality of facilities and equipment

- quality of science
- quality of service
- understanding of the customer's requirements
- handling of the contract
- pricing
- value for money
- effective use of the client's time
- delivery on time and on budget
- handling of complaints
- meeting the client's requirements
- communication of the results
- overall satisfaction
- rating of other competing research providers
- willingness to use the organisation again.

Like all forms of market research, CVA needs to be conducted regularly to be of most value and to pick up trends in the relationship. The drawback for scientific organisations is the cost of doing this sort of research frequently – including client annoyance.

To gain an all-round view of one's science, it is highly desirable to compare the findings of CVA with other forms of public opinion research, especially with quantitative and qualitative research among opinion leaders and among consumers or taxpayers. This is the only way you can be completely confident the science is acceptable to society.

REPUTATIONAL ANALYSIS

Reputational analysis (RA) provides a way for companies and organisations to measure the good standing of their brand in the wider community and with particular groups of stakeholders.

RA is an evolution of brand valuation, which in its original form was focused chiefly on monetary value (as expressed by turnover, share price, etc.) but which grew to embody a range of other less tangible, though no less important, values affecting overall corporate performance such as social and environmental approval.

In the 'triple-bottom-line' era, when the performance of companies and institutions is judged by social, environmental and ethical criteria as well as fiscal results, RA is becoming quite popular with larger corporations. It reflects an understanding that the financial outcome can be seriously affected by poor performance in, say, the environmental area or by an appearance of unethical or insensitive behaviour. The same applies to research institutions, whose funding depends not only upon having strong science projects and good staff but also upon the image they project for trustworthiness and level of attunement to community views and values.

In its fuller forms, RA is detailed and very expensive – probably beyond the means of most scientific organisations. Nevertheless, it its simpler form it is a handy way to assess the reputation of the research organisation in the eyes of the public, not for reasons of corporate ego, but to assess how much confidence and trust the public will have when asked to adopt or use its science.

The RA process seeks to:

- identify and analyse factors that contribute to the institution's public reputation, good or poor
- identify stakeholder groups and opinion leaders who are important in the establishment of a sound reputation
- analyse current and future risk factors that may affect the organisation's good standing with various stakeholders
- create a 'reputation profile' that lets the organisation measure its ability to meet stakeholder standards and expectations.

Some forms of RA allow an organisation to track its standing directly against its competitors and see who's ahead in the image stakes. However, many institutions must take part in the survey for this to be informative. Others place the emphasis on measuring the organisation's reputational performance according to criteria it has set itself. This is probably the more useful, as it lets a body judge whether it is succeeding or failing by the light of its own ideals.

Like other forms of opinion research, RA involves putting questions to influential individuals and organisations across the community – such as peak industry councils, environmental bodies, religious organisations, government institutions, professional groups, and so on. The answers are collected and analysed and may be consolidated, if desired, in a reputation index that can move up and down just like a financial credit rating. The kinds of issues that can be measured include:

- value to customers
- value to the community
- value to funders and investors
- quality of management
- social impact
- environmental impact

- ethical standards
- external communication
- staff relations.

The potential benefits of RA for science include greater external investment, more partners, more rapid and effective adoption of results, greater public transparency and trust, and greater attractiveness to scientific talent.

An important feature of RA is posing the same questions to staff internally. This reveals whether there are major gaps and inconsistencies between self-image and external image. For example, the staff may on balance regard their outfit as highly ethical but the wider community, reacting to one or two unsavoury events that were widely publicised, may award a far lower score. When such perceptions flow into reduced research funding or lower uptake of an organisation's knowledge output, it is in trouble.

READING THE PUBLIC MIND

A new variant of customer value analysis known as 'reading the public mind' has been introduced by one of the authors and colleagues – as a way of discovering how the public is likely to react to the introduction of a potentially controversial and powerful new science or technology, and in the hope of avoiding future cases of rejection of the sort that occurred in the case of GM food in many countries.

This uses a continual sample of public opinion gathered via the internet to creating a 'moving picture' of what the public thinks about the issue at any time. Based on a powerful statistical technique developed by Dr Nick Fisher, it explores not only whether the public is for or against a particular issue or technology, but also the reasons for its views – and how these change over time. This provides a level of clarity about public opinion that is simply unavailable using 'snapshot' opinion polling or focus groups. Being internet based, it is also a great deal cheaper.

For the first time 'reading the public mind' enables the scientist or institution to understand the many factors, pro and con, which go to forming public attitudes to a particular science or technology, on a continuous basis. This provides powerful information that can help the research institution decide whether it should:

- educate and inform the public more about the new technology
- change the technology to conform to public concerns or wishes, or
- drop it altogether, as it is unlikely ever to be approved and is wasting precious funds and time, and move on to something more likely to be useful and acceptable.

Also, for the first time, 'reading the public mind' offers science something that it has lacked throughout almost its entire history: data about how society will receive new knowledge or a new advance. Science itself is founded on data, yet most scientific institutions employ only guesswork when it comes to assessing whether or not their outcomes will be acceptable to society or widely adopted. 'Reading the public mind' takes away some of the guesswork and replaces it with testable data.[2]

Although other effective methods for communicating with stakeholders exist – in particular the creation of standing advisory committees – the ones described above share the virtues of being relatively inexpensive and effective ways of gathering opinion widely from across society. However, the best opinion research in the world is of scant value unless the organisation is willing to listen.

The most important strategic ingredient in a 21st century scientific organisation is the ability to listen to the outside world – to clients and stakeholders, but also to large audiences including government and the community, minorities, native peoples, regional and interest groups.

The future of science depends on its ability to shape itself to the needs, values and standards of humanity.

The future of humanity depends on a science that is open, listening and committed to overcoming the inequities and inequalities caused by the uneven sharing of knowledge or its misapplication.

Chapter 5

Communicating with the media

Like science, the media is about ideas. It is a natural forum for the discussion and debate of new scientific findings, and their dissemination and acceptance by society. To the journalist, science is an inexhaustible source of news – not only about discoveries, but also about the application and meaning of science for society, and the inevitable controversies that surround these. In open science, journalists and scientists are thus partners in the sharing of knowledge, although they do not always view themselves in this way.

What hinders the partnership on so many occasions is the retention of stereotypes. To journalists, the scientific archetype is the wire-haired male boffin with the mad gleam, the ratty clothing, the strange equipment and incomprehensible vocabulary. To the researcher, the journalist sometimes appears to be a wolverine, red in tooth and claw, jamming a foot in the lab door preparatory to dragging that scientist's reputation through the mud before the scandalised gaze of their colleagues and the world at large. Like all stereotypes, these fail the test of genuine experience, yet many in both professions cling to them, and it is the communicator's job to overcome them.

Nonetheless, scientists and journalists inhabit very different cultures and observe contrasting imperatives, as Table 5.1 suggests.

There are many social, economic and environmental gains to be made from a stronger partnership between science and the media. Science is about the creation of knowledge and exploring ideas. The media is about sharing, debating and testing ideas in society's marketplace. The great majority of people gain almost all they know about new science and

Table 5.1 Differing needs of scientists and journalists

Scientists prefer:	Journalists prefer:
• detail, data, method	• application: what it means to people
• to be rational, cool and objective	• emotion and drama
• teamwork and shared credit	• heroes, not teams
• incremental progress	• 'breakthroughs' (hot news)
• to qualify their views	• controversy/conflict
• to consult peers	• clear, crisp comment NOW!

technology from the media. A strong partnership between science and the media is therefore essential to an advancing society and to open science.

By transmitting new knowledge, or at least the awareness that new knowledge exists, the media helps people to improve their lives more rapidly and effectively. This empowers individuals. It offers a means for tackling poverty, ignorance and disempowerment because it gives people access to the information they can use to take charge of and enhance their own lives. It helps governments, industry and society's leaders to make better-informed decisions and to change behaviour or technology that is found to be damaging to society and the environment.

Virtually nothing significant in science is free from media scrutiny at some point, especially at the point of application within society. And there is little in science that does not actually profit from this scrutiny, whether it is simply the community being made aware of the advance, helping to find a commercial partner or assisting a government to formulate sound policy. Moreover, the media can make scientists aware of unforeseen consequences of their research – whether it has potential for wrongful application, is out of step with social and moral values or whether it can be enhanced by being presented in a different technological package. The media lubricates the successful uptake of new knowledge, and helps society move more rapidly and with greater agility towards sustainability, economic and social progress.

In most countries, the media is the main way that older people learn about new things. However, the media does not see its job as providing society with a classroom science education, or even bringing about a fundamental improvement in national scientific literacy, other than in the broadest sense. It is not about drumming up new recruits for science (or any other profession). Mostly, the media focuses on the application, meaning and impact of science to society and ordinary people or on its entertaining and fascinating aspects, rather than on the technical detail of research. Because of this, some scientists regard the media as an obstacle – or at least an inconvenience – to their work, and this view prevents them from taking full advantage of mass communication tools for spreading new ideas and ways of thinking.

In recent years, people in modern democracies have begun to insist on their right to be informed about new ideas and new technologies before they decide to accept, adopt and use them. When it comes to new technologies, people wish to ask the tough questions, listen to and be persuaded by the answers. They usually seek objectivity from science – not advocacy.

They wish to witness the debate, and be assured that their main concerns about safety, ethics or the environment, as well as cost and practicality, have been addressed or at least taken into consideration. They are becoming increasingly intolerant of the old, patronising 'science knows what's best for you' model.

Humans are understandably risk averse. We have spent the last 3 or 4 million years in the development of a marvellously sophisticated system for identifying, confronting and limiting the dangers that surround us. It is one of the secrets our evolutionary success. This is the main reason why the media often seems to be full of bad news: not because journalists and editors like it that way, but because readers and audiences demand it and the media who ignore this market imperative soon go out of business. Finding out about danger is a human survival trait. People want to assess the risks so they can set in train the social mechanisms to neutralise them. Scientists are particular beneficiaries of this process: they are often paid by society to do research that reduces risks, and that helps make our world a safer, cleaner, healthier and greener place. However, society also wishes to be reassured that new solutions do not themselves contain graver risks.

Risks generally first come to public attention through the media, and their repeated presence in the media is a clear signal to politicians that it is time to act. The political response is frequently to direct more resources to science (and other areas) in order to minimise the perceived risk. Astute scientists frequently take advantage of this. So, next time you wonder at the media's apparently insatiable and gruesome appetite for food scares, cancer threats, pollution hazards, accidents, plagues, crime rates, climatic shifts, fires, crashes, floods, mortality rates and daily disasters, one way to view it is as more work for science. Another is as the working of the unquenchable human instinct to survive.

There is a feedback loop between science and society, much of which is provided and lubricated by the media, which enables the ideas, concerns, criticism and debate to pass to and fro. A few scientists – intimately familiar with every aspect of their specialised field – are impatient of this process, hankering for the days when authoritarian regimes simply decided what was good for the people and the people did as they were told. Today's smart scientist understands that working through the media is an intrinsic part of delivering the benefits of science – and avoiding some of its downsides.

In the 21st century we have entered the age of the democratisation of science, when people not only demand a say in the outcome and how it is used, but in the very science itself and how it is performed. It is an age when good science is judged not only by its scientific quality, but also by its social acceptability and its value (or harm) to ordinary people. It is an age when people are (or should be) equal partners in the development and adoption of new knowledge and technologies.

The media is therefore a primary target for science because it reaches all the other audiences – decision makers, opinion leaders, professionals, industry, partners, competitors and the community at large.

ENSURING ACCURACY

An error often made by people with little media experience is to regard the journalist as the target of their message, whereas in fact the journalist is only a channel through which the message must pass in order to reach decision makers or the wider public. How accurate that message turns out to be depends significantly on how much effort was spent on ensuring accuracy in the first place. When approached by the media for comment, there is a tendency for researchers to 'wing it': to do the interview off the cuff, without much forethought or the preparation of clear background material. Then they are amazed and outraged when the story turns out cockeyed.

The good news is that this does not need to happen. To a very significant degree, the researcher or scientific institution can help the media to 'get it right'. With care and forethought, they can make the media their partner in delivering a scientific message to the wider community. However, as in all human relationships, the key ingredient for this to occur is trust.

THE IMPORTANCE OF TRUST

The journalist must be able to trust the scientist that what they are being told is truthful, correct and not self-interested, because the journalist almost never has the time, specialist knowledge or resources to validate a scientific claim objectively. They have to take science on trust.

The scientist must be able to trust the journalist that their work will be fairly and truthfully reported, without exaggeration, distortion or misrepresentation.

Such a relationship of mutual trust does not spring up during a single interview. It requires time, repeated contact, and understanding of each other's needs and aims. It profits from contact at an informal, human level as well as the professional level. Where this happens, the relationship can be extremely fruitful for both parties. Scientists can cultivate relationships with journalists who deliver their findings accurately to society – and journalists can in turn learn to rely on particular scientists or institutions who deliver quality, unbiased research results, on time and in a way that can be easily understood by a lay person.

COLLATERAL BENEFITS

Accurate science reporting not only brings the scientist's work to the attention of policy makers and the public, it also brings it to people wishing to fund research or invest in its commercialisation. And, unexpectedly, it can bring it to the attention of other scientists working in related fields: a growing number of scientific collaborations, especially international, begin with the partners becoming aware of one another through a report in the general media.

A valuable but intangible benefit of good science reporting is the boost it gives to the morale of the research team, arising from public recognition and social validation of their work. Sometimes, too, experienced journalists can contribute real value to science from the breadth of their experience by suggesting ways for its adoption that are likely to be more politically or socially acceptable, or unexpected applications in unrelated fields. By asking the right questions, they can teach both scientists and institutions how to frame their explanations of complex science for a general audience.

The journalist who cultivates scientific contacts over time discovers an ever-refreshed fount of news stories. He or she also benefits from reliable tip-offs received from researchers about issues that at first may seem to hold minor significance but which grow to assume national, even global, importance. For the journalist, regular contact with scientific institutions and their staff produces the things that further a reporter's career – exclusive news breaks, broader contacts, a wider information base, insightful analysis, clues to new directions in which society may move and a deeper understand of the world we inhabit.

The journalist and the scientist are partners in the process of knowledge generation and sharing. Each can add a wider value to the other's work.

UNDERSTANDING JOURNALISTS

Journalists are as varied a group of individuals as scientists, many of them just as bright and, occasionally, brighter. Usually they are a worldly bunch, with a well-honed sense of how changes or discoveries are likely to be received by society, or by sections of it.

As a rule:

- Journalists are mostly in a hurry, with a deadline to meet. They want to get straight to the point and not beat around the bush.
- Journalists are inquisitive, probing and shrewd in how they obtain information. They are trained to uncover secrets and to detect untruths. It pays to be prepared, tactful, open and polite in dealing with them.
- The curiosity of a good journalist knows no bounds. It will often outrun the scientific data available, but nevertheless requires answers.
- Their focus is on their reader or audience and what these will make of the story (though it may also be on the editor and what they interpret as the interests of their audience).
- A journalist is neither the inferior nor the superior of a scientist – both serve the society in different and special ways.
- Journalists deal with the high and low in society. They are less impressed by rank, status, position, honorifics, awards, prizes and qualifications than other people, and more by genuine human qualities and abilities. They often have highly developed ways of summing people up.
- They work for money and the media is mostly in business to make money. A science story will often be assessed in terms of its significance to the public and its capacity to generate an audience for the media. The size of that audience governs the income of the media, because advertisers pay to reach particular audiences with precision.
- Like a good scientist, a good journalist is cool, detached and objective in dealing with the information they gather. They are rarely there to praise or condemn, but to report and, sometimes, to analyse.
- Journalists are interested in people as well as facts. The media thrives on human-interest stories within and around the news. They seek

the human or emotional side of a story because that is what their audience is also interested in.

- Journalists don't enjoy being made a fool of, or being treated with disdain any more than do scientists. However, because of their access to the public, they can do a lot of damage to any person or institution that offends them or whom they decide is arrogant or contemptuous towards the public interest.
- The scientist and the journalist are equal partners in a free society and in the process of sharing knowledge. The relationship should be one of trust, mutual respect and collaboration.

GIVE THE BACKGROUND

There are two kinds of journalists: generalists, who may be reporting on a fire or flood one moment, a social event the next, a movie star after that and a law court or political rally later on – and specialists, who cover a particular area such as politics, economics, business, medicine or science. When dealing with a general reporter, it is important to take the time and trouble to fully explain the background of the research to be communicated. No prior knowledge should be assumed on the journalist's part. Even if the journalist does have some understanding of the issue, their audience probably does not – and the message must be designed for them, not for the reporter. Help the journalist to interpret the meaning of the work to society.

A common mistake made by scientists is to assume that the reason they are doing a particular piece of research is self-evident. It usually isn't. In an interview, it is very important to answer the question 'Why are you doing this?' – and the answer must be couched in terms of societal, not scientific, benefit (to save lives, protect the environment, make industry more productive or competitive, create jobs, etc.).

Good, clear information leads to accurate reporting. Keep the facts as uncomplicated as possible. Use plain language and simple concepts and examples. As discussed in Chapter 2, avoid specialist jargon, especially terms that have several possible meanings – one to a branch of science and a different one to the community or other professions. It is vital to bear in mind that it is not the journalist the scientist is addressing, but the public, in all their wisdom and ignorance, with all their prejudices, preconceptions and language difficulties. The message must be shaped so its meaning is clear to them, whoever they may be. Scientists sometimes object to this on the grounds it is 'dumbing down' their science, but the opposite

– communicating science in ways that cannot be understood or are open to misinterpretation by the audience – is not very smart either. There is almost always a middle way by which the essential message of the science can be transmitted, though it may take hard work and clear thinking to hammer it out.

As a rule, specialist journalists are far more knowledgeable in their field. As they are usually senior reporters with wider experience, they will provide better coverage in terms of accuracy and quality of treatment. They also have more influence with the editor in achieving extensive coverage and better placement of the story. It pays to cultivate these journalists as regular, reliable contacts.

UNDERSTANDING 'NEWS' AND WHAT THE MEDIA WANTS

Unfortunately, in our media-saturated age, some scientists despise the general media and do not take much trouble to read, listen, watch or understand it. They sometimes insist the daily media should adopt the same editorial standards as *Nature* or *Science* – a treatment that the vast majority of humanity would find indigestible and inapplicable, if not incomprehensible. This attitude can cut science off from the society it serves and make it harder for scientists to know and provide what the media, or society, expects from them.

The professional science communicator must therefore read the main daily papers and monitor the internet, TV and radio news and current affairs programs continually, serving as the 'eyes and ears' of the scientific organisation, helping scientists and their managers to understand how and when to pitch their stories and how to interpret what is being reported. This often makes the communicator the bearer of unpleasant tidings – for example, having to explain to a biotechnology institute why society criticises and rejects its offerings. Conversely, the communicator can often spot opportunities to deliver scientific messages or new technologies that have opened up as a result of events reported by the media – so enhancing both science and institution in the eyes of public and stakeholders.

Familiarity with what the media sees as news is critical to effective science communication. A research institution that can couple its work and its achievements with the 'news' is sending a clear signal to society and decision makers that it is in tune with events, relevant to current the needs of the society and deserving of more support and greater funding. On the other hand, an institution that seeks to promote its work at a time when the news focus is strongly elsewhere risks being seen as irrelevant, out of touch, self-seeking or an 'ivory tower'.

The understanding of what constitutes 'news' is vital in helping to structure and time important announcements. It is also essential in identifying opportunities to 'hook' the science story to running news, and so emphasise its relevance to issues of current national, international and local importance.

News, like beauty, is in the eye of the beholder, and different media define news according to their own priorities. However, for science, it can be defined as:

- something that is completely new, today
- something that has just been found to have a big impact – monetary, social or environmental– on society
- something said or done by somebody important, famous or notorious
- something odd, quirky, amusing, shocking or bizarre that is not generally known
- something that contradicts a popular belief or policy
- a new publication, report, release of figures
- something of importance or interest to the particular audience addressed by specialist media
- a natural disaster, accident or event and its consequences
- something controversial or scandalous
- something entertaining or liable to be talked about.

Scientific discoveries and advances match many of these categories. There is therefore a natural 'fit' between science and news, of which the good communicator and astute scientific leader or researcher takes full advantage.

DEVELOPING CONTACTS

The media should be closely studied for reporters whose treatment of scientific issues suggests they have an interest in your kind of research and the skills to report it well. Build and maintain a list of contacts to whom you speak regularly. Contact them at a personal level, as well as a professional one, so that the relationship grows without a sense of one exploiting the other. From time to time, offer them information that is not available to other media, because there is no more convincing argument a journalist can use to their editor than 'We've got an exclusive here'. This plays to one of the media's greatest strengths – its competitiveness. A newspaper or electronic medium with an exclusive story is strongly inclined to run it before its competitors can, and to give it prominent treatment. Once the news breaks, the competition is then strongly motivated to try to 'top' the story – to better it by developing a new 'angle' that allows it to catch up – so it is also sensible have follow-up information ready and waiting.

However, in providing exclusives, avoid being seen to play favourites among individual journalists or media, as this can antagonise the others. As a rule of thumb, 'big' stories should be released at the same time to all major media. Lesser stories in terms of news value can be released to individual media as 'exclusives', but it is wise to rotate one's favours among key media. It is also advisable to match the science stories to the particular strengths of different media – a good medical research story, for example, is best delivered to the medium with the best health coverage or reporter. This recognises the skills of a particular media outlet, gives them encouragement to continue their good coverage and, at the same time, places competitive pressure on the rest to lift their game.

Meet the reporter personally soon after you make first make contact. Seek to get them to the laboratory or to a field research site where you can build an effective working relationship and trust. This also allows the journalist to check out the 'visuals' – opportunities for photographic or TV imagery that go with the story – and to size up the 'talent': the ability of the scientist or research partner to deliver the message in a way that meets the need of the medium, whether print, radio or TV. It also allows the journalist to see the scientist in context, surrounded by their tools of trade, and gauge their enthusiasm: these provide the atmosphere that helps create a strong, well-rounded story.

PROVIDING 'MEDIA FACTS'

When providing material for the media, cost–benefit or social benefit figures on the research are essential. To journalists, these are the core facts on which they will often hang the story.

Why was the research done in the first place? What are the likely benefits to society or the economy of its adoption? What is the size of the global or national market for this technology? How many lives could it save or benefit? What measurable difference might it make to society or to the environment? How much did it cost to perform the research? Who paid? Why?

The media dotes on dollars and cents. They are one way of translating the significance of science (and other things) into terms that ordinary people relate to. (Hence the media's eternally irritating question about priceless works of art: 'How much is it worth?') Scientists are sometimes offended by the conversion of an elegant insight into the crudity of cold cash but, objectively, it is a way of helping society to value it against countless alternative activities that taxpayers are asked to fund. It should be seen not as belittling the science, but rather as an indicator of its significance to the wider community and a justification for investing in it.

In describing a scientific advance, accentuate the benefits to consumers, to taxpayers, to urban (or rural) society, to jobs, to health, to export income and to the nation. Even if the story is about pure, 'blue sky' or theoretical research, its relevance can still be conveyed to a general audience by explaining how, one day, its application may change their lives or improve their standard of living in certain ways. Although this offends the ideal of discovering knowledge for its own sake, it nevertheless helps account for why taxpayers have helped fund it, which is fair enough if more funds are to be sought in future. It is also a reason to at least explain some basic aspects – though not the detail – of commercially secret research, if it has been assisted with public funding.

THE MEDIA IS INSATIABLE

In the media, there is always another news bulletin due, the printing press will be hungry again tomorrow and the web page must be refreshed to attract traffic. All journalists are continually in search of a story – either a new story or a new slant on an old story – so don't feel hesitant about ringing a specialist reporter at work or home, day or night, on weekends or public holidays with a tip-off or lead. They will be especially appreciative of fresh information that reaches them on a 'slow' news day, such as a weekend or public holiday. These are good times to schedule media announcements about science, as there is less direct competition for space from other kinds of news.

The media rarely sees a science story as being about science; it is usually as a story about politics or industry or society that happens to have a scientific ingredient. Thus there are many places for a science story in the media – from the nightly news bulletin to a gardening show, or the sports, entertainment, IT or arts pages. The main places to locate a science story are:

- in 'general news' (at the front of the paper or on the TV or radio news bulletin)
- in 'features', which are longer in-depth articles or magazine-style stories of about 800–2500 words, usually accompanied by pictures or in TV and radio current affairs and discussion shows

- in special sections such as the medical section, the health section, the education section, the IT section, the car pages or home pages, which are generally supported by advertising
- in regular sections such as sport, the arts, business, entertainment and weather
- in 'special reports', which are sometimes one-off advertising supplements
- in programs about people
- in radio discussions and talkback
- in letters to the editor and opinion editorials or opinion programs on radio and TV
- in international news and current affairs sections and programs
- in specialist science programs and sections (though judiciously, as these tend to be watched or read by a minority audience with a particular science interest).

DEADLINES

Whether they work for electronic, daily, weekly or monthly media, all journalists face unrelenting deadlines and their professional reputation stands or falls on their ability to deliver a complete and accurate story on time. In radio, TV or internet news, deadlines may occur several times a day. On weekly or monthly publications, with fewer staff, deadlines are less frequent but often even more demanding and stressful for reporters than in the daily media where there is plenty of material to choose from and far more staff to cover it.

> *Respect for a journalist's deadline is one of the best ways to build a sound relationship. Failure to respect deadlines is also a way in which communicators and scientists who have never experienced the pressures, stress and drama of a real newsroom damage their own reliability as contacts.*

If a journalist seeks information and it is not immediately available, promise to get back to them at the earliest opportunity. This means *minutes*, not hours or days. Promptness in dealing with media requests is a critical element in the ability of a scientific institution to establish and maintain a reputation for communicativeness, cooperation and openness.

When a journalist decides to work on a story, they notify the news editor and the story is placed on the news list, along with all the other possible stories for that edition or bulletin. An expectation builds up among the senior editors that the story will appear, fully researched, prepared and checked, and by the deadline or earlier. If, through a failure of the journalist's contact to reply, the story does not appear, the journalist is seen to have let the news team down. They failed to deliver. No matter how plausible their excuse, this harms their standing with their editors in an intangible way. Several such events have the editors questioning whether that particular journalist can deliver the goods. So a journalist who is let down by a particular contact or organisation usually avoids them in future. They are likely to transfer their focus to other contacts: other scientists who have shown themselves willing and able to provide information in a timely way.

A reliable, deadline-conscious contact is one that journalists will use time and again. A casual or unreliable contact soon drops off their list. Although individual scientists might not care too much about this, in today's competitive world the bottom line may be a decrease in funding for their work or institution, due to the loss of its public profile and perceptions that it is no longer doing such worthwhile work. Thus a poor attitude to the media can have an impact on the public standing of the organisation, its funding, and hence its future scientific success rate and ability to recruit talent.

INTERVIEWS

If a scientist has been asked for an interview by a paper, a TV or radio program, the interviewee needs to take the time and trouble to prepare their comments, develop the background information and identify the two or three salient points they wish to make. A communicator can help greatly in this process by adding the external perspective on how the interview will appear to the media and public, as distinct from the potentially myopic internal view.

Nowadays, a typical commercial TV news story is 30–80 seconds in length, radio even shorter and a print news story about 250–500 words. This means that the 'live quotes' from the scientist will take up only about 8–15 seconds of the electronic story or a few sentences in the print version. Asking a scientist to sum up several years' work in ten seconds is probably not possible. The interview will therefore depend on that single crisp, concise and colourful sentence that explains for the audience *what the science means to them.* Such sentences seldom slip off the tongue in the heat of an interview. They usually have to be crafted, worked and re-worked with exquisite care beforehand. Then they have to be rehearsed. Finally, they have to be repeated during the interview, sometimes several times in slightly different ways, to give the media a choice of useable quotes. This may seem a nuisance to the busy researcher, but the extra care will be rewarded by better media coverage and fewer mistakes.

Radio offers one of the friendliest ways to accustom researchers to meeting the needs of the media, especially those mid-morning, mid-afternoon or evening guest interview shows that tolerate a bit of rambling and have time to delve into how the science was done. Here the interview may last for 5, 10 or even 15 minutes, and eases the scientist gently into the process of shaping and editing their story for the media. They soon learn, from the reaction of the interviewer, where the interest lies, and what is considered 'boring' to the radio audience. They learn, too, that as human beings, their feelings and personality are just as fascinating as their discoveries. Many a scientist has been stumped when describing their great breakthrough by the simple question 'Yes, but what exactly did you *feel* when you made this discovery?' At heart, the media wants every discovery to be heralded by the naked elation of a 'Eureka!'

As part of the preparation for an interview, it is important to ask beforehand:

- What is the interview about (i.e. is it really within the interviewee's expertise, or is it likely to stray into political shoal water)?
- How many minutes or seconds will it be on air? This gives a good idea how to martial the argument and whether there is time to make more than a single key point.
- Who else is the interviewer speaking to on the topic? This indicates whether they are trying to set up a controversy or debate and whether it is about the actual science or something else.

It is sound practice for the interviewee to jot down their key points to make sure they are all covered in the time allotted. As experience grows, the interviewee learns how to turn almost any question back to the topic they wish to pursue.

In handling a controversial issue it is important for the communicator to carefully select the spokesperson on the basis of experience, presence of mind and clarity of delivery. At the same time, suggest spokespeople from other organisations who can support your claims or statements. This saves the media time hunting around for corroborative evidence and reinforces your own story. It also provides them with variety in the form of a wider range of people to quote.

ON AND OFF THE RECORD

When speaking with a journalist, every interviewee or communicator should be aware they are speaking to the public, however private they might imagine their confidences to be. Under the journalists' ethical code, a reporter is technically obliged to caution you if you are 'on the record' (being reported), but in practice this is usually assumed by both parties if the media has contacted you seeking comment. It is sensible to assume you are on record at all times in any dealing with the media, unless it is specifically agreed that the discussion is confidential.

It is important to spell out exactly whether what you are saying is intended to be reported (i.e. *on* the record) or offered purely as unattributable background (i.e. *off* the record). Always feel free to ask 'Am I being quoted?' If the journalist replies affirmatively, then ask to have your quotes read back to you, so you can clarify anything you may inadvertently have told them in the innocent belief that you were 'just chatting' or off the record. If giving an electronic interview, remember that one of the oldest tricks in the book is to leave the recorder running after the interview has apparently ended, while the journalist broaches the sensitive issue apparently out of an innocent personal interest.

In dealing with a sensitive issue, in which you may wish to make remarks about other individuals or bodies that you wouldn't want to see linked with your name in cold print or on air, the best tactic is to say 'First I'd like to give you the background, which is off the record … afterwards, I'll give you some directly attributable quotes'. Then clearly distinguish the point when you move from unattributable to attributable.

Never tell a journalist 'You can't report this'. If it's good news copy or in the public interest, a journalist can report anything they want, within the bounds of the law. If you don't want something reported, don't tell the media.

These points all underline the importance of building a relationship of trust between journalist and scientist or science communicator, where each comes to rely on the integrity of the other and there is cooperation rather than a fear of exploitation. Effective science communicators often use the media by providing background briefing to their trusted contacts, who can then write accurate stories attributed to 'informed sources' or 'a leading scientist'. This is a good way to raise issues that governments, bureaucracies, corporations and other elites don't want exposed, but that need to be aired in the wider public interest.

In recent years, considerable pressure has been brought to bear on scientists in many countries working on the public payroll not to comment on issues such as climate change, food safety, pollution, fish stocks or forestry issues because it may be inconvenient to governments or industry to have these concerns aired. However, one of the most important functions of science is to inform humanity objectively, and without the taint of politics or vested interest, what is going on in the world that the public needs to know about. It is in this regard that the 'off record' briefing is most commonly used, though those who employ it should be aware they may be targeted in a witch-hunt for whoever 'leaked' the information and should take reasonable precautions (such as not using the office email!).

TROUBLESHOOTING

Politeness and courtesy are keys to developing good relations with the media, even if a reporter annoys you through ignorance, persistence or having got something wrong. The thing to remember is that it's not the reporter you're serving, it's science. It will pay you to get the message over patiently, clearly, politely and – if necessary – repetitively.

Never use the phrase 'No comment': it's like taking the US 5th Amendment on the grounds it may incriminate the speaker. It makes any journalist smell a rat. Answer every question possible, but if you have to duck one, then explain carefully your reasons for doing so (e.g. 'I can't answer that because it is covered by a legal confidentiality clause, but I can give you a general idea what the work is about ... '). People who say 'No comment' are often regarded by the media as arrogant towards the public interest and as having something to hide. They will soon find themselves under scrutiny.

When misreported, the immediate reaction is to get on the phone and yell at the journalist. Don't do it. Go and have a cup of coffee, calm down, then come back, and ring to see if you can correct the mistake and turn it into extra coverage by persuading the media to do a follow-up report that clarifies the situation.

The reason for this is that it may not be the journalist's error in the first place – and you will never know it. In a newspaper, for example, the mistake could have been made by another journalist whose copy was incorporated into the same story, by the copy sub-editor, the page sub-editor, the news editor, the chief of staff, the editor or any of a dozen different journalists who handled the writer's story after it was filed. Blaming the journalist will only antagonise them and lose you a useful contact. Above all, don't blame the journalist for the headline, which was almost certainly written by a sub-editor and is designed to grab reader attention – not to summarise the story.

However, the origin of the mistake may also lie with the scientist. Perhaps they didn't take enough care explaining their story in the first place?

Besides giving an interview, did they also provide a plain-language document explaining the work? This may seem like a chore, but often an existing document such as the summary from a grant application or the abstract from a scientific paper will help. Better still, write a simple plain-language summary of the work and its meaning and make sure the journalist leaves with it in their hand. When they are back at their office and find something in their notes they can't understand, a glance at the summary may clarify it. It will also correct spelling. It should provide contact phone numbers for the scientist, not just in the office but also on mobile and after hours, because the journalist may still be at their desk, filing copy, long after the scientist's working day has ended.

Some scientists ask to view the finished copy to ensure accuracy. This is a step that must be taken with extreme care, and only by a scientist with some experience in dealing with the media. The reason is that it infringes a basic journalistic principle of impartial, objective reporting, free from external influence. The scientist's intent may simply be to ensure scientific precision, but, for the inexperienced, the temptation will be strong to tinker with how the journalist reports the story – and that can rapidly lead to a fight. As a rule, a scientist may offer to vet the story and their direct quotes for scientific accuracy. They cannot insist. They do not own the story, even though it is about their work, and the journalist is under no obligation to show it to them. Good journalists, in our experience, do not usually object to someone helping them to 'get it right', but they strongly resent any attempt to change the way they are reporting it, or any delay that may cause them to miss deadline. The journalist knows far better than the scientist what their editor requires and what their outlet will publish or broadcast. With the best of intentions, the scientist may so distort the news value of the story as to make it not worth running. Then everyone has wasted their time and the journalist is unlikely to return to that scientist or institution.

One reason for such misunderstandings is the contrasting publishing traditions of researchers and journalists. News reporters concentrate the significance of their story for their audience in the first paragraph – the 'lead' – then support the initial claim in the paragraphs that follow. Scientists, on the other hand, prefer to build their argument on a bedrock of methodology, data and prior research, before reaching a conclusion. Some researchers seem to delight in concealing their conclusions in paragraph 89, thus making the reader work all through the argument to reach it. Media style is based on newspaper practice, which allows the reader to browse the headlines and lead paragraphs in search of the information they desire. The importance of the story to the reader emerges at once, in the first few paragraphs. The 'lead' is designed to sell the story to the reader, to catch their eye, to make them want to read it. The traditional writing styles of scientist and journalist are thus almost opposites. A scientist who seeks to impose scientific style on a news story will usually only succeed in killing it.

WHAT NOT TO DO ...

As a matter of good practice in dealing with the media, don't:

- demand retractions
- make threats
- run to the lawyers to issue writs
- abuse people who may be quite innocent
- complain to a journalist's editor unless you are sure they are guilty of a breach of professional standards
- write abusive or hysterical letters that will only make you look foolish.

All of these can get the media offside, and can make a bitter (and unnecessary) antagonist, with a memory like an elephant and a capacity to harm or ridicule your organisation over time out of all proportion to your ability to strike back. In the case of an error or misreport, a cool head is essential. It is better to try to amend the situation by generating an accurate follow-up story either in the offending medium or else its competitors than to force the media to back down and admit error, as this often yields confrontation.

A considered response to a misreport involves careful analysis of who actually saw the report, and who may have believed it. The impact of a misreport cannot, and must not, be judged by 'coffee-room indignation level' or the blood pressure of senior management. The science communicator often has to spend a fair amount of time hosing down internal outrage before planning a measured response. The correct strategy is to work out which audiences saw the report, whether or not they believed it and, if so, how to deliver an accurate version to them with precision. In devising a plan, it is important not to draw the dispute to the attention of a much wider audience and more media, which may only escalate it and cause the error to be reported more widely.

Many institutions shoot themselves in the foot by over-reacting to a minor report on the inside pages of the paper and either attacking the media or demanding an unnecessarily prominent correction. Often, the effect of this is to turn the scientific organisation's purple-faced outrage into the subject of the next story – and to advertise its thin skin to the rest of the media as an open invitation to journalistic mischief. In many cases, wisdom will decree it is better simply to write the error off, provided it is unlikely to cause material harm to the scientific and public standing of the organisation. It'll mostly be forgotten by tomorrow, in the hectic news world.

If, however, it is judged that key audiences and the public have been misled and there is need for a correction, a measured, temperate letter to the editor stating the facts is one way to set the record straight. A press release can work, but may on occasion serve to escalate the dispute because it will engage the offending media's competition. Press conferences should only be called in the most serious cases of misrepresentation, and the response strategy very carefully thought out. The law is the last resort of all, if only because a writ represents a declaration of war and the law is not, in any case, the best forum for resolving scientific disputes. The guiding principle in all such disagreements should be the public interest, as distinct from injured organisational ego.

Scientists often fear a media misreport because of the effect they perceive it may have on their science, their standing with their colleagues and their future career. However, experience shows the risks are much lower than their imagination may sometimes depict. In one scientific institution, young scientists who had never had a media experience were asked what they thought the outcome would be: five out of six were sure they would be misreported and made to look foolish. When scientists who frequently used the media were asked the same question, five out of six said the experience was usually either satisfactory or directly beneficial to their work. The one story in six that went wrong was regarded philosophically as the price of doing business, and seldom as an irretrievable negative.

If you do encounter a genuinely prejudiced and hostile journalist, who is shoving a toe in the door and demanding answers, it is important to respond with courtesy and polite promises to get back as soon as you have verified the situation. It appears far better on TV than a guilty-looking figure scurrying into the lab for safety.

To sum up, in the event of a media error or misreport:

1. Estimate how many people may have seen the error and what effect it will have on them. This should guide your response and where to direct it.
2. Match the response to the size of audience and target them alone. Overkill will only blow up the issue and widen media interest in it, leading to the misreport being repeated elsewhere.

3. Match the response to the need of the audience for correct information.
4. Seek a follow-up story that clarifies or amends the mistake. Try to turn a negative into a plus by working with the media, if they are in a cooperative frame of mind.
5. Demands for retractions or legal threats will antagonise the media, invite closer scrutiny and won't help you get your message over. They may blow up the issue and become the story.
6. Don't immediately blame the journalist for the error. It could have been committed by any of the many others involved in the story, including you. Try to find out what went wrong.
7. If the issue is determined to be damaging, consider writing a cool, rational and short letter to the media body concerned, amending the error. Avoid abuse, blame-laying, sarcasm, and so on, as these reflect on you and your institution.
8. Don't issue a media release in response unless absolutely essential – it can inflame the issue and encourage other media to take an interest in it, not necessarily in your favour.
9. Don't blame the journalist if the story does not appear – most stories fail to get into the daily news. Seek a new angle and another opportunity on a different day.
10. If an individual journalist is proving negative, consider various ways of getting them on side rather than attacking them – which is what they may want you to do.
11. Consider not responding at all but letting the mistake 'go through to the keeper'. This is often the wisest strategy, as it will soon be forgotten in the busy daily news cycle.

WHY YOUR STORY DIDN'T GET IN ...

Don't blame the journalist if your story failed to make it into the newspaper or electronic bulletin, or if it is cut very short. It is almost certainly not their fault. On a big newspaper there may be 50–100 writers or more, each filing one or more stories and comment pieces every day, plus there are scores more coming in from contributors or news services. Out of this pool of 200 or 300 possible stories, the editors may choose the 25–35 stories they think are most worthy of a run. The other 80–90 per cent end up 'on the spike'. The same applies, and more so, to TV, radio and internet news sites. The pressure and the competition for newspaper and electronic news bulletin space are enormous.

If you story didn't get in, it is probably because, stacked against the news of the day, it didn't rate strongly enough in the opinion of the editors, no matter how important you thought it to be. Remember, your story is seldom competing against other science news – but against general news about politics, the economy, world developments, wars, crises, accidents, unemployment, business, crime, sport, the arts, celebrity capers, and so on. There may have been nothing wrong with your story – it just wasn't as strong as the competition it was up against at the time the news editor received it and compiled their news list.

This is why a science news story must be as hard, vital and as relevant to the audience as possible. Its release should be timed for a day when general news-flow may be down or there is a running news story connected to the same issue.

If your story fails to get in, try to find out why and analyse the reasons. Some of them may be within your power to influence, and you can arrange things better next time. One major reason a science story fails to make it is that it is considered 'boring'. This does not mean it lacks intrinsic interest, but that it has no obvious relevance, immediacy, colour or human interest to connect it to the media's audience. Mostly, this means not enough time was spent considering how the media would respond to it, or how to engage the public's interest. Or else that it was so changed and diluted by the internal approval process that it finally emerged devoid of any news value.

Media releases particularly misfire when the institution is more concerned with trying to burnish its image than it is with the true significance of its work to society. The media has plenty of time for the latter, but a very limited tolerance of the former. Stories perceived as having more to do with corporate ego or self-justification are, rightly, considered unpublishable.

PUBLIC FIGURES

Well-known public figures can to help promulgate the outcomes of research – as well as coffee, shampoo or charitable causes. If the story itself is a bit dry in media terms, it can be enlivened by having a prominent or colourful identity linked with it. Politicians, industry leaders, actors, media personalities and sporting heroes can all be employed in this role. It works best if they have some personal connection to, or interest in, the research.

Although many scientists will recoil at the mere thought of having a rock star help publicise their work, the media and general public will be fascinated to know why the celebrity is interested in being involved. The late Princess Diana's global promotion of the land mine issue is an example of the effectiveness of this technique, and many movie stars have been associated with emergency relief aid and charitable causes. Al Gore's promotion of climate change science is a relevant example, as is David Attenborough's promotion of conservation science. Qualities that celebrities can bring to science awareness include:

- greatly enhanced photographic, TV or interview potential for the media
- the 'curiosity factor' of why the celebrity is involved
- public endorsement of the importance of the work
- the representation of the work in terms that ordinary people can more readily relate to, by virtue of the celebrity's involvement
- improved scope for fund raising or investment in R&D.

CONTROVERSY

Controversy and debate are what make democracy tick. They are the way we test important issues and decisions in the market place of ideas – and the media adores them. However, many researchers – especially older ones – eschew public controversy (though they may be perfectly keen to engage in vigorous debate within their peer group). As a result, their science receives a lower external profile and their advice is sought less often, perhaps, that that of more clamorous parts of society. Controversy should be used to advantage in open science, rather than being feared – although this must be done with skill, careful planning and some effort to anticipate the various consequences.

A scientific organisation keen to demonstrate the value and relevance of its work and share its knowledge must be prepared at any moment to engage in national or global debate on sensitive issues. Too many institutions content themselves with merely issuing statements on matters *they* regard as important, while ignoring the opportunity to participate in the larger debates that the public, politicians and media regard as important. Too many scientific institutions practise self-censorship, rather than speak out about matters liable to make politicians, industry or lobby groups uncomfortable. Developing sound policy to deal with this is discussed in Chapter 10.

If a scientific body wishes to contribute more effectively to the national dialogue, its leaders and communicators should scour the media daily for issues on which they can comment expertly, interestingly and with insight. Don't wait for the media to come to you – go out and look for them.

MEDIA TRAINING

A scientific organisation serious about sharing its knowledge, contributing to national dialogue, building its profile and attracting research funding, will ensure media training for key spokespeople. These include not only senior managers and directors, but also the leaders and deputy leaders of scientific research programs and, in some cases, much younger scientists who are seen as having the right skills for media work – particularly if the organisation wishes to project a youthful, vibrant or female image to counter the 'aged male boffin' stereotype.

Good media training usually involves exposing the researcher to working journalists from the different media in a lifelike interview situation. Replaying interviews and having the journalist explain where the scientist 'went wrong' is very instructive. However, it is important not to pitch the training at too threatening a level, as the aim is build confidence, not to undermine it.

Working journalists can be a great help in identifying who is, and isn't, good media 'talent'. One of the critical jobs of the communicator is to know whom to put before the media and whom to keep right away from them – the abrasive, the arrogant or dogmatic, the impatient or those who simply do not grasp the function of media in a modern democracy. Being a brilliant researcher does not always equip a person for the role of science communicator.

Media training usually takes place at three levels:

- *introductory*, to help the scientist understand the needs of the different media, how to satisfy them and how best to put their work across to achieve accurate reporting
- *medium*, for senior researchers and research leaders who have had some media experience, but are now starting to find their work in the public spotlight or subject to growing controversy. This provides a higher level of skill and confidence in handling tricky questions
- *advanced*, for chief executives, directors of research, deans, vice-chancellors and others who may find themselves subject to aggressive media questioning in a crisis situation where responses need to be handled with speed, tact, consistency and skill.

Very few scientific organisations invest in a regular program of media training for their staff, which is an oversight as it neglects the main tool for delivering science to society. Those that do are the organisations that recognise that effective communication is what gives true value to science

– and this requires understanding to be developed on both sides, not just on the part of the public and media.

MEDIA RELEASES

Releases are an effective way of communicating a scientific announcement to the media. They have the advantage of being circulated far and wide, of presenting the facts in a coherent and informative way, of providing essential details and contact points for media to follow up, and of being checked and cleared by all concerned. They put your version of the story 'on the record'. A good science media release consists of:

- an eye-catching headline designed to appeal to journalists (*not* to scientists or institutional egos)
- an opening paragraph stating the significance of the announcement to the general public or a target audience (e.g. an industry or policy makers)
- details of who is making the announcement, with correct titles
- text written in news style, with one sentence to a paragraph, one idea to a sentence, and the most important information at the top
- lively quotations attributed to the lead spokesperson to give the media a sense of the colour and importance of the story
- recognition of all relevant partners and collaborators in the work. Note that this should go lower in the media release. No matter how important these institutions may think they are, their names (usually, alas, long, cumbersome and opaque) are of secondary significance to the media
- clear statements of the importance of the work to the economy, jobs, society, ordinary people, industry, government policy, particular regions or localities, or the world
- a simple explanation of how the science was done and how it works (a diagram or graphic can help here)
- contact details for spokespeople at work and after hours, and for a communicator who can locate them within minutes if need be, along with email and web addresses
- details of picture or vision opportunities for TV and news photographers
- a web address for other still pictures (in jpg or other format), sound or vision that the media can download. Even a Facebook site, if the science is important.

News desks and journalists are engulfed in a daily avalanche of paper and electronic information. Many hundreds of announcements must be evaluated for newsworthiness and priority in a short time and reporting resources allocated. Most of these announcements receive barely half-a-second's scrutiny from a fast-reading editor, chief-of-staff or news reporter before being deleted. A strong, eye-catching headline combined with a powerful, crisp and relevant opening will save your release from such a fate.

Make your releases snappy, concise, newsy, factual and relevant to the external audience. Make sure the news is in the headline and the first paragraph. Purge it of all jargon. Don't bury the main point or it may be overlooked.

Ernest Hemingway trained as a journalist on the *Kansas City Star*, and always recalled the injunction in its style book:

> *Use short sentences. Use short paragraphs. Use vigorous English, not forgetting to strive for smoothness. Be positive, not negative.*[1]

The exhortation applies equally to the well-crafted science media release (and is, in some respects, the antithesis of traditional scientific prose). To this is added the necessity to avoid clichés, tautologies, ambiguities and specialist terminology, bureaucratic language or 'technospeak'. Here is a checklist of things to consider in issuing a media release:

1. Does the text answer the six key journalist's questions: Who? What? When? Where? How? And most importantly, why?
2. Who checks it? Did all the research partners see it? Who gives final clearance for it to go out?
3. Which media or journalists should be targeted for optimum impact?
4. When should it go out? What else is happening in the global news stream that may conflict?
5. Should it be distributed locally, nationally or internationally?
6. What is the best mode of delivery: email, fax, hard copy, web or conference?
7. What picture, vision or sound opportunities must be planned?
8. What graphics or multimedia must be prepared in advance?
9. Is the spokesperson freely available for a day or more after the release is issued? (Some scientists have a curious habit of issuing a release and then disappearing.)
10. Can the spokesperson be easily reached out of office and after hours?

11. Is the release 'stand alone', or can it be 'hooked' to major topics now running in the news, thus increasing its prospect of coverage?
12. Which key journalists should be notified in advance that a release is planned to ensure they receive it?
13. If the topic is controversial, have the relevant stakeholders and partners been fully briefed?
14. Is there a follow-up story? Is it better to plan a series of releases to build awareness over time rather than issuing a single 'fire-and-forget' announcement?

EMBARGOES

If you don't want your announcement published at once, but to await a special time or event such as a press conference, public ceremony or publication date of a scientific paper, then put an embargo on the release requesting the media not to publish the content until the specified date and hour. Most news media will respect this, although it is wise not to expect too much of a highly competitive industry in the case of a really big story.

Put prominently on the first page the words 'Embargoed until', followed by the desired date and time of release. Check carefully the deadline times of your target media so as not to select an inconvenient time or, worse, one that falls after their deadline has passed. Avoid, wherever possible, setting an embargo time that favours one media outlet or medium over another, as this can anger the loser. An embargo may be for several hours or several days ahead of the release time. In view of the international 24-hour reach of the media, especially via the internet, you may have to make clear the embargo time in several key time zones such as GMT or US eastern and western time.

This technique is especially useful for releasing substantial material that is used to provide background to important announcements, and most journalists will be highly appreciative of the extra time it gives them to assess the information, work on it, arrange for vision, interviews, graphic art, and so on. The result is more accurate and fuller coverage than would normally be the case.

TARGET MEDIA

Careful thought should always be given to targeting media announcements. Many releases are wasted because they are poorly designed for the particular needs of the media they are hitting, and only serve to annoy the journalists who have to sort and discard them.

Different media require different kinds of stories or different treatments ('angles') of the same story. Sometimes it pays to make several slightly different releases on the same topic, each one tuned to the needs of a distinct media group.

Examples of media categories requiring different treatment include:

- national daily press and electronic media
- business and industry newspapers, magazines and sections
- feature sections in major papers
- current affairs programs on the electronic media
- specialist publications for business and industry
- computer and IT publications and sections
- medical and health publications and programs
- food, diet, nutrition and cooking media
- lifestyle media
- sporting media or sports sections
- rural and farming publications
- regional and local media
- women's, family and general interest magazines
- radio and TV chat shows.

A well-crafted general release may meet the needs of all the main target groups, but one that is precisely targeted at a particular part of the media will, on the whole, enjoy fuller coverage and be more effective at reaching the desired audience. A sensible approach is to develop a general release and then rewrite just the headline and leading paragraphs to appeal to different media categories. A technology story, for instance, can be presented in terms of its overall significance to society for general media ('New drug to save lives'), its financial importance for business media ('New drug to boost corporate earnings') and its parochial significance for local media ('New drug brings jobs to Smithtown').

Another way to do this is to plan a series of announcements over several days, weeks or even months, each time targeting a different media category or else presenting a fresh angle on the same story. Science shuns repetition of the same old data or theory, but it is absolutely essential in public awareness. The art lies in telling the same story many times, but always with a novel and distinctive twist, at an advantageous time and to a different audience. At one moment the focus may be on the scientific advance, the next on its significance to humanity, then in its commercial consequences, then on the people who achieved it, and so on. Modern knowledge management theory contends that the 'story' is in fact an ideal way to share knowledge. Homer understood this 2700 years ago, journalists have known it for at least three centuries, and now giant technology corporations are also discovering that stories can be used to transfer knowledge informally. It is a fact that most people prefer *stories* about science, rather than having to digest plain, unadorned data and excruciating prose.

To reach different media audiences with precision and impact requires a list of media contacts sorted into different audience categories, and an effective electronic distribution system. This will usually consist of a compilation of email addresses and fax numbers for key media organisations and individual journalists. Public relations companies, telecommunication firms and media organisations provide commercial media release distribution services. Although laborious, it may be better (and cheaper in the long run) to assemble one's own unique media contacts list and maintain it so that it is always up to date. Media directories can be purchased for most countries, and even for the world, although they too are often out of date. Nevertheless, they provide a starting point for assembling a media contacts database that can then be supplemented by direct calls to the media itself.[2]

The most effective way to use such a database is to plan every release in terms of its individual target audience(s) rather than using the scattergun approach and risking irritation to receivers who do not want it. The release can be sent to particular categories of media and journalists (e.g. farming media, health media, science reporters or environment reporters) or even to individuals within each category. Because the media tends to have a strong local focus, it is very useful to be able to target a story at a particular region, city or locality affected by it.

It is sensible to send duplicate copies of the same release not only to the key journalist, but also to their chief-of-staff or news desk. This ensures the editors are aware of the story, even if the journalist is away. It helps to place it in the news system.

Some stories lend themselves to targeting news media, while others suit the more relaxed setting of radio or TV talk shows, feature sections and magazines. It is valuable to be able to distinguish between the different categories when distributing a release. As a rule, a news story contains material that is new to the media and has never been published in quite that form, whereas feature material focuses more on the wider context, in-depth treatment or human interest aspects.

MEDIA CONFERENCES

Media conferences are held to make general announcements to a wide range of media, or to a specific group of journalists (e.g. science, medicine or environment reporters).

These are a good way to make important announcements, but beware: they can backfire if not well-managed or if the subject is not sufficiently significant in the eyes of the media. As a rule, a media conference should *only* be called if the topic is of major importance to the external world. It is not something to be staged on a corporate whim, but should be used sparingly and with professional judgement.

Failure to observe this fundamental rule can result in a spectacular disaster if the media become sufficiently frustrated and annoyed. They, after all, make a major investment of staff, time, equipment and money to attend, in the expectation that they will obtain a strong and newsworthy story. If the conference fails to live up to their expectations or to meet their needs, at best the journalists and their managers will be reluctant to devote time or resources to covering future events by the offending institution. At worst, they may become aggressive.

The following are some basic rules for organising a successful media conference.

1. Ensure there is a genuine news story to be announced, with strong local, national or global significance. If in doubt, take professional advice from a friendly working journalist.
2. Field your best spokespeople at the conference, ensuring they are well briefed beforehand, know exactly what to say and are confident performers for TV and radio.
3. Time your conference carefully to allow all the target media adequate time to digest, report and analyse your announcement before their deadline. As a rule, a morning conference is preferable to one held in the afternoon or evening, when media deadlines mostly fall.
4. Issue an alert or invitation, both to key journalists and to their news desk, several days ahead of the event, to ensure your conference is in their news diary. Try to avoid a clash with other major media events (sometimes a friendly chief of staff or news editor will advise you of this).
5. Provide a detailed media release or background kit to all journalists who attend, and to others who may not be able to come. Make sure it contains all relevant facts in plain language, as well as digital illustrations or video the media can use to clarify the subject to their audience.
6. Keep opening statements short and to the point. Don't waste time duplicating information that is already in the kit, but focus on the key announcements and the 'grabs' (the quotes you hope the media will use).

7. Plan the 'punchline' to the conference well in advance. This point will be made, in different ways, by the various spokespeople taking part. Allocate other key points among the speakers.
8. Provide spokespeople from partner institutions, industry or government to confirm the value of what is being announced.
9. Provide picture and film opportunities for TV and print photographers including, where possible, attractive imagery such as intriguing scientific equipment (*not* computers), research settings involving animals, people or plants, and picturesque locations.
10. Rehearse your spokespeople ahead of time and train them not to use technical jargon and not to say 'no comment'.
11. Provide time and a place for one-on-one interviews with the key spokespeople for individual journalists afterwards. Remember that each journalist may want to obtain a degree of exclusivity in what they report.
12. Ensure your spokespeople are freely available to journalists for follow-up interviews after the press conference, as they may wish to check facts prior to their deadline. Provide working and after-hours contacts.

Chapter 6

Communicating with government and industry

GOVERNMENT

Scientists and politicians may seem an ill-assorted team, yet they are essential partners in the process of sharing knowledge with society.

Most scientific institutions depend on public support from governments and legislators, either as direct funding or through incentives that encourage private sector or philanthropic investment. They also depend on governments to incorporate the findings of research into public policy for the benefit of the whole community. At a pragmatic level, about three-quarters of all new patents lodged worldwide begin life in a publicly funded research institution, even though they pass into private ownership subsequently.

In the face of these facts, one would expect the relationship between scientific institutions and the world of politics and government to be close. It is therefore a surprise to find that scientific institutions have limited skills when it comes to understanding and dealing with governments and politicians. Also, they devote relatively few resources and scant effort to furthering the relationship. Where science does make attempts to influence the political process, its efforts are usually unsophisticated and fragmented, at least when contrasted with those of industry, professional lobbyists, non-government bodies and the like.

As funders and users of science, ministers, members of parliament and the senior public servants who advise them are among a research organisation's most important clients. They, after all, hold the key to how science influences national policy and the effect it has on the nation's progress.

What MPs think about science

Scientists are sometimes shocked or offended to learn that science is a low-order issue with politicians. It seldom provides the substance of major political debates and, when it does, is quite often resented by politicians for having intruded without warning into their world. The key to an effective partnership between science and politics is for scientists and their managers to develop a better insight into the way the political mind and process works, and then deliver their research findings according to its needs. In dealing with politics, a 'science of service' attitude works rather better than the traditional 'we're here to give you the scientific facts' approach.

Politicians of all parties and backgrounds generally understand the need for research, and very few politicians are actively anti-science.[1] In surveys, most Members of Parliament rate science as either important or very important. Statements such as this, however, are akin to endorsing motherhood and do not convert automatically into increased support for R&D, or the adoption of scientific findings in policy.

When it comes to funding, science often seems to the busy MP just one in a long queue of worthy causes demanding money. The scientist pleading for increased resources for something they passionately believe in may be unaware that they were just one of dozens of supplicants who importuned the minister or MP that week. In the eyes of their advocates, *all* these causes are urgent, all are *vital* to society, all demand national priority, they affect the peace of the realm or the health of the community, and they have social, monetary and political benefits such as higher employment, greater income or environmental sustainability ... and many cater to marginal electorates. The politician's job is to reach an informed – often intuitive – decision about which causes to back, and which they can afford to ignore. (Issues that appear in the media or in public opinion surveys are much harder to ignore.)

Politicians have differing views about the adequacy of scientific research funding. In one Australian survey, 52 per cent of federal and state MPs considered that science was inadequately funded, whereas 48 per cent either had no opinion or thought it to be funded sufficiently. This was in a context in which the objective evidence pointed to a real decline in national science investment over decades, suggesting the science lobby had not got its message across. A third of the MPs admitted they had never actually considered the question of whether science was adequately funded. Although the figures may vary from country to country, it is likely there exists in all legislatures a proportion of representatives for whom science is not on the political radar.[2]

Another intriguing revelation is that not all legislators grasp the link between science funding and economic or security outcomes, as the following quotation from an Australian MP disturbingly illustrates:

> *Science and technology are important. But scientists also need to understand that economic prosperity and defence probably rank higher.*

This MP clearly regards science expenditure as a budget drain, not as an investment in the economy, and technological leadership as having little to do with defence. Such is democracy. However, there is in fact a political basis for this MP's position – although it is one few scientists would appreciate: any science this MP helps fund today is unlikely to have a significant pay-off in their political lifetime, or to help them to be re-elected in a year or so, whereas an industry subsidy, welfare handout or tax cut has an immediate pay-off. The scientist may deplore such short-termism but it is as much a reality in political life as the law of gravity is in physics, and science has to learn to work with it.

Budget priorities

When times are tight, extra money for science can only be found by making cuts elsewhere in the budget – in health care, defence, education, law enforcement, and so on. This is a practical reality that science advocates sometimes overlook. While making a good case for extra funding for research, they fail to offer practical suggestions about where the money might come from. This omission may convey to politicians an impression that scientists imagine that money grows on trees.

For the MP, the issue is how and when to trade off perceived political pain against political gain – what will be the cost in disaffected voters from the cuts to, say, defence, education or social welfare, compared with the increased electoral support likely to flow from greater research funding. This simple political calculus explains why the going can be tough for science, compared with other fields – because it lacks a strong public constituency to applaud the government's decision to back it. It also explains why medical researchers often do better in the budget stakes than their colleagues in other fields – saving lives is a highly visible political justification, whereas the benefits of a better grasp of string theory are a bit harder to sell to the electorate.

This does not make it impossible to mount a case for enhanced funding of theoretical and blue-sky science. However, it does mean that the research organisations seeking it need to be smarter and more imaginative in how they go about it. In some cases, they can link their work to clear national goals and political priorities. In others, the mere promise of a piece of expensive equipment – a synchrotron or radio telescope, for example – will do the trick because of the visible kudos it brings to the politicians who backed it.

However, an important key to securing their favours is providing MPs with detailed cost–benefit and social benefit analyses of the work in question, allowing them to exercise their political judgement about its relevance and importance and its political 'saleability' to the electorate. They also need clear and simple examples of the likely benefits or outcomes of the research, so they can explain and justify it to the people.

A slightly different approach is needed in dealing with the bureaucracy, where science proposals may come under the chilly scrutiny of treasury and finance department officials. Here the key measure is return on public investment rather than political instinct or public approval. Marie Keir,[3] a former senior government advisor to CSIRO, argues that case studies demonstrating how much the nation, or industry, actually gains from certain research are far more persuasive than heartfelt appeals such as 'we *must* have a scientifically literate population' or 'we *must* keep up or risk being left behind'.

The most influential advisers, she says, are usually those at senior executive level of the public service. These are the people asked to comment on submissions for science funding, and their attitude towards the applicant may be of great importance.

Competitive funding ranks high with bureaucrats because it seems to them to imply financial rigour and efficiency (which is not always true, alas, given the paperwork involved). But it also puts some distance between them and the responsibility for actual decisions. Proposals that mesh with national priorities and government policy goals are also likely to win bureaucratic approval as well as political support.

Lack of contact

One of the most damning remarks made of modern science is the occasionally voiced complaint of politicians:

The only time I ever see a scientist is when they want money.

An Australian survey found that only 18 per cent of federal and state MPs regarded the interaction between scientists and politicians as 'successful'. The rest considered it to be neutral (42%) or unsuccessful (40%).[4] If only one MP in five is happy with what they are getting from science, then scientists clearly have a lot of political homework to do.

The reasons given by the MPs for their dissatisfaction include:

- lack of cost–benefit data on the science
- lack of face-to-face engagement
- scientists' belief that government should fund most research
- inadequate communication of scientific issues and findings through the media.

One-third of MPs said they never met a scientist during a typical year, while a further 38 per cent said they received some form of scientific briefing only once or twice a year. This indicated that 71 per cent, or nearly three-quarters, of Australian federal and state MPs had little or no direct contact with either science or scientists.[5] Something similar, no doubt, applies in most countries.

This may seem bad enough for science, but there is an even more alarming aspect: at any one time – to use the Australian Parliament as a case study – this cohort of science-deprived MPs will include four and possibly six future prime ministers and more than 100 future ministers, who will thus have little exposure to science and technology in the course of their political careers, but who will come to wield enormous influence over it. They will be called on to assess the outcomes of scientific research for relevance to public policy, and to rate them against less rational claims for their political practicality and affordability.

As might be expected, those MPs who never or rarely come into contact with science also hold the lowest opinions of its importance, the need to fund it, its policy usefulness and its ability to meet politicians' needs. A good many of these politicians belong to the numberless class in the community who were 'put off science at school', regarding it as arid, dull and the province of nerds. It is distressing, but not hard, to see how the prejudices of adolescence may be reflected in national leadership and policy decisions decades later.

The mismatch in timeframes

The timeframes of politics are short: the present 24 hours, the next seven days and the approaching election are critical punctuation marks. The timeframes of science, technology and innovation are long: results are delivered over years, decades, even a generation or two – long after the government that originally funded them has become history. MPs understandably see few immediate political pay-offs from a decision to bolster science funding, or even from a decision not to cut it.

On the other hand, science is seldom professionally equipped to deliver results or information with the immediacy, clarity, simplicity and political relevance demanded by MPs. It is ill-accustomed to translating its outcomes into present-day or immediate future social and economic benefits with which politicians may wish to associate themselves.

Yet this is what science has to do if it wishes to influence the political system. It needs to explain its activities in terms of the political, not the scientific, timetable. Politicians, more than any group, are conscious of the fragility of promises: to them a scientific promise is no more reliable than a political one. They need to see it being delivered here and now, or at least in the very near future – not in '5 or 10 years'. They'd like to take some credit for it, or share a bit of reflected glory. They would like to see science actively improving the lives and jobs of their electors, preferably before the next election is called.

Scientists, of course, are forever scrutinising the future, often the distant future. So much so, in fact, that they may have lost interest in the work they did 3, 4 or 5 years ago, and that is at last starting to be adopted by society or industry. Consequently, they omit to communicate to politicians the *present* benefits of *yesterday's* work. Sound science communication ensures that the process of knowledge sharing goes on long after the actual R&D has wound up, and keeps current benefits of old research fresh in the public and political mind.

Electoral 'clout'

The number of people employed in research and development in any country is seldom more than 1 per cent of the voting population and, in developing countries, a great deal less. Researchers are generally distributed fairly evenly across electorates, which means that science and technology issues rarely influence the election outcome in a particular constituency or determine the future of an individual MP, let alone the fate of governments.

From a political standpoint, science thus lacks political 'clout'. It has nowhere near the local influence of a major industry or even a big factory, or the national influence of a sector of the economy or major group in the community.

A second reason for this lack of electoral clout is that scientific institutions seldom form alliances with industries or groups to whom politicians pay attention. The academic desire to remain uncontaminated by the grubbier aspects of industry or special interest lobbying is understandable.

However, this should not prevent research institutions from joining forces with powerful lobby groups on particular matters of national public interest, or from encouraging industry to lobby on behalf of science. A call for increased science funding is far more credible and cogent coming from a captain of industry than from a scientific leader.

Political 'scientific literacy'

An Australian study of MPs in 2001 found only 16.5 per cent of MPs and Senators had professional qualifications in science, technology, agriculture, engineering or medicine. Fewer than 6 per cent of all MPs held science degrees.[6] These levels are likely to be mirrored in other legislatures, especially in developing countries. Not only do most politicians have scant familiarity with how science and innovation systems function but, as previously noted, they do not always automatically connect them with issues such as employment, wealth generation, security or sustainability. However, scientific literacy among MPs and ministers, although helpful, is not an absolute requirement for the adoption of scientifically sound policy or of sound science policy. It is more a question of the quality of the relationship between scientists and legislators, and of the character of the communication between them.

In his analysis of the shortcomings of science in communicating with its various audiences, Roederer[7] puts the boot firmly on the scientific foot. He says:

> *We are witnessing an alarming erosion of public trust and political support of science and knowledge-generating institutions, to the point that some outright anti-scientific threads have become evident in many parts of popular thinking.*
>
> *I believe that to a large extent we have to blame ourselves for these problems! I believe we scientists indeed are naive and socially ineffective – maybe reasonably good communicators in the classroom, but generally bad communicators with the public, the media and the politicians.*
>
> *Indeed our greatest threat may not be the scientific illiteracy of the public, but the political illiteracy of scientists!*

The cultural divide

To politicians, perception – what most of the public appears to believe at the time – is the reality within which they operate, regardless of personal conviction or scientific 'facts'. Scientists deal in measurable data, whereas politicians mostly rely on experience and intuition, fuelled by opinion polls, to forecast how an issue will play out.

Scientists often find it hard to accept the apparently irrational forces of popular belief and prejudice that sway democracy. They are uncomfortable operating in a world in which perceptions, rather than facts, predominate. Indeed, researchers not infrequently voice disparaging opinions of politics and politicians on this account, which are unhelpful in the process of building a working relationship. There is a tendency to overlook the fact that the whole of democracy is a giant negotiation and a series of trade-offs, in which facts may be less significant than opinions, beliefs and the concessions that must be made to them.

Because of this, many scientists are inclined to shun engagement in the public policy debate out of a fear of having their objective findings weighed in the same scale as popular prejudice, or out of distaste for the horse-trading that accompanies every major decision. To the politician, however, such reticence may look like a lack of confidence in the science. In any event, this shyness will decrease the likelihood of the science being quickly adopted.

However, professional cost–benefit and social benefit analysis can help to bridge the gap.

The linguistic divide

Scientists speak specialised languages without realising how pompous, opaque or excluding their words may sound to others, including politicians. Politicians, too, have a language that, to scientific ears, sounds curiously selective and evasive. Asked in one survey, 'How is science travelling at the moment?' 84 per cent of MPs responded: 'Not well'. When asked why, some said lack of funds but most blamed it on poor communication by scientists.[8]

Poor communication refers directly to the use of jargon and not meeting face to face, but it also refers to scientists' failure to translate their arguments into terms critically relevant to a political audience: the effect on the economy, on jobs, on public opinion, on various vocal interest groups, and on current high-profile political issues.

If you ask a scientist to explain the impact of their research on public opinion, you are likely to encounter a resentful stare. Yet this is precisely the sort of information politicians need when they evaluate the political risks and advantages of various courses of action advocated by scientists.

Throughout this book, we argue that open science demands early and constant feedback between science and society, with knowledge travelling in both directions. It requires the scientist to acknowledge that the community, too, is a possessor of knowledge about its own needs, values and preferences, which can complement and assist the knowledge developed by research. If this takes place, not only is the ultimate technology likely to enjoy a smoother path to adoption or commercialisation, but the scientist no longer needs shrug when asked by a politician what the community is likely to make of their findings.

In addition to speaking eloquent physics, chemistry and biology, today's scientist also needs to be able to speak politics.

Improving the science/politics relationship

For all the foregoing reasons, science and support for science are rarely regarded by politicians as essential to political success, or among the highest national priorities.

Scientists are far from alone in feeling they get a rough deal from politics. Many industries, professions and community groups consider themselves short-changed by the political system: farmers, miners, business, non-government organisations, welfare agencies, unions, environmentalists, sport and the arts, to name a few. Like science, these groups share a conviction that *their* activity, in particular, is central to the nation and its future.

A salient difference between science and other groups is that the others have mostly adopted modern, professional, well-planned approaches to the task of influencing policy and persuading decision makers. The scientific approach is decades behind the times, which is a little strange considering how futuristic the orientation of science is in other respects. As one MP put it:

> *The science lobby has to become more ruthless and persuasive to convince politicians that science should be backed.*[9]

A 'professional' approach to government does not mean the hiring of high-priced lobbyists or PR firms, although a little advice from them on what works and what doesn't can help.

The secret is for scientific institutions to learn to put themselves in politicians' shoes, to see the world through political spectacles or, bluntly, to understand their customers' needs. This does not require every scientist to become a *quasi* politician. But it does mean having skilled individuals on the team who know how the political mind works and are across current issues and events, and science leaders who are comfortable and experienced at working with politicians. And it means having one's case couched in political rather than scientific terms, backed up with straightforward and professional cost–benefit and social benefit data.

To bowdlerise J F Kennedy, recruiting politicians to the cause of science involves asking not what politics can do for science, but rather what science can do for politics.

What MPs want from science

Research by Australia's CSIRO has looked at how MPs obtain information about science and technology, and how they use it.[10] A federal MP focus group said their main interest in, and use for, scientific information lay in the areas of:

- the current political agenda
- current public concerns
- global developments
- new and emerging issues
- important scientific advances
- current research projects.

They indicated that they mainly obtain scientific information from:

- the Parliamentary Library
- the web and email
- organised science briefings
- personal briefings, as requested
- committee briefings
- visits to research sites
- the media and newsletters
- their advisers
- CD-ROM and video (low).

The MPs said they used scientific information for:

- helping with policy development
- internal party discussion and debate
- material for speeches in public and in Parliament
- to counter incorrect information by interest groups
- to promote investment in science.

Their main requirements were that scientific information should be:

- timely
- relevant to current political issues

- in plain language
- re-useable by MPs
- accessible after hours
- from a credible source
- provided at various levels of detail
- with executive summaries
- with case studies and 'stories'.

Other key elements stressed by MPs as helpful in building a partnership between science and politics were:

- fast access to the right scientific expert, a directory or point of contact
- a face-to-face relationship based on mutual benefit, not a series of one-off contacts with random scientists hustling for funds
- scientific information packaged in small, regular and digestible chunks rather than huge, confusing dollops or long, jargon-laden reports
- use of concrete examples to explain science, instead of abstractions and generalisations
- use of the 'story' technique to describe what happens in science, and engage the listener or reader
- personal visits to labs and scientific sites to generate insight, enthusiasm and personal contact.

The stand-out factor here is that politicians all want scientific advice about the current political hot issue *now*. They don't want advice about yesterday's hot issue, or even tomorrow's (except in rare circumstances), and certainly not next year's. Scientific institutions are generally ill-prepared to satisfy these instant demands for information, yet their performance and value to the nation is being subtly appraised by politicians based on whether they can do so.

In reality this is not such a tall order as it might appear, as the political demand is mostly for simplified and background information rather than intense detail. The maintenance of a comprehensive file of issues briefs – in paper and electronic form and compiled by someone who is well-informed about what is going on in the legislature and what is coming up and can write lucidly – will usually suffice. The other ingredient is a reliable 'ear to the ground' in the Parliament or bureaucracy, which can feed back information about emerging issues on which a scientific opinion or input will be of value.

Fostering the dialogue

Dr Gary Gray, in his career both a politician and a science manager, offered the following rules for improving the dialogue between scientists and politicians.[11]

1. Research is about wealth generation and the future, not just science. Focus on things like jobs, improvements in living standards and a cleaner environment.
2. Engage politicians and community leaders locally. Inform them and help them to understand the issues.
3. It will take time to create a culture of understanding for science in politics. Don't expect a single meeting or a quick fix.
4. Science can win by steadily building a case for science.
5. Articulate outcomes, and what the science means, in plain language.
6. Be an enthusiast for all science, not just a lobbyist for your own discipline or institution.
7. Build support methodically. Start by inviting MPs to visit your institute.
8. Be conscious of the political cycle.
9. Talk to all sides, including minor parties. Remember that today's opposition is tomorrow's government and today's backbencher is tomorrow's cabinet minister.
10. Be informed about the legislative process.
11. Be useful to politicians. Don't raise problems unless you have solutions to offer.
12. Keep dialogue simple and factual. Substantiate your claims from reputable sources.

Gray stressed the importance of building a dialogue between scientists and politicians as the primary requirement for developing a stronger scientific culture in politics. It is not good enough, he says, simply to demand money for research. He urged scientists to avoid berating politicians, but instead to try to win them through an infectious enthusiasm.

'Remember,' he advises scientists, *'when you meet with a politician, you are no longer only a scientist. You, too, are a politician'.*

Similar advice came from US physics professor Juan Roederer. He counsels scientists to do their homework first: find out the politician's specific responsibilities, political views and personal interests. He urges them to prepare a case using simple metaphors or examples the politician will relate to – and not improvise during the talk. He also suggests keeping key points and data on cue cards for ready reference, remembering that politicians are great debaters, ready to leap on any inconsistency.

Roederer also has an excellent list of 'nevers' for speaking with MPs:

- never talk about yourself unless asked
- never mention money (funding) unless asked
- never contradict a politician ever if you disagree with him or her
- never make a statement the politician may interpret as a threat
- never use acronyms or scientific jargon
- never hand scientific or technical papers to the politician
- never take up too much time
- never raise unrealistic expectations about what science can deliver or what it predicts will happen
- never appear condescending.[12]

Building a comprehensive relationship

The links between a scientific institution and government may occur at many different levels and in varying ways. Typically, the points of contact consist of:

- the ministers for science and the relevant portfolios
- Cabinet submissions and requested policy advice
- senior staff and advisers to ministers
- senior public servants and departments with an interest in science, and their committees
- Parliamentary policy committees, on request
- government scientific advisory groups
- individual members of the legislature and their staff, on request
- policy committees of political parties, on request
- the political media
- parliamentary information sources, such as the library and information network.

The following are some tested ways that links between science and politics can profitably be enhanced.

National science briefings

Special science briefings have been staged in the Australian and Indonesian federal parliaments and Australian state parliaments to provide politicians and their advisers with up-to-date scientific insights into topical issues. The ideal briefing involves two or three speakers from different institutions or backgrounds, including industry, each limited to 10 minutes and using the latest presentation aids (e.g. PowerPoint, multimedia or hands-on demonstrations). Speakers are carefully coached, purged of jargon, focused on the political (as distinct from scientific) needs, and their notes made available afterwards on paper, by email and web and from the Parliamentary Library.[13]

The aim is to show politicians (and their advisers) how science can assist better policy decisions. It is to promote the value of science to politics. It is *not* to brag about a scientific institution or lobby for funds – which would undermine the aim of the briefing. It is to send a message that scientific knowledge can underpin sound and durable policy.

Briefings need to be supported or sponsored by all the main national scientific organisations to avoid jealousies and the sort of infighting that discredit science when compared with other, more organised and disciplined industry or community lobbies. It is helpful if the Science Minister or a very senior politician hosts them.

Similar briefings can also be staged in state and provincial legislatures. Here the accent is on how science can contribute to local and regional development, to the progress of particular local industries and to help tackle local environmental issues. Speakers include leaders of local industry or the community to give politicians a strong sense that the science being advocated is also politically sound in the local electorate.

Science updates for electorates

This email service is precision-tailored to the needs and interests of individual politicians. It notifies them, in brief, of scientific discoveries and outcomes directly affecting their local constituency.

It is based on a database containing information on the main industries and sources of employment in each electorate and the personal interests of individual politicians. For example, a story about an advance in cattle nutrition might go to politicians with a dairy or beef farming industry in their electorate, while an advance in breast cancer screening goes to politicians with a particular interest in women's health issues.

Because politicians are inundated with paper and information, the email is not sent directly to them. Instead it goes to their electoral adviser. This important person manages the MP and constantly briefs them about the state of local opinion, local issues and local concerns and what has to be done and said as they move around the electorate.

The electoral adviser uses the science update to give their politicians something to talk about as they address community or industry groups, give interviews to local media or write their columns and newsletters. It is a subtle way of enlisting the politician as a messenger and advocate for science.

Discoveries in big national research labs seldom receive local media coverage for the simple reason that they are national, not local. However, the inclusion of the local MP's name magically transmutes the science into a local story, so this is another way to increase general media coverage of R&D.

'Science meets Parliament'

Over the years this has been a particularly successful experiment by the Federation of Australian Scientific and Technological Societies (FASTS) to bring scientists and politicians into face-to-face contact. It involves scientists leaving their familiar turf and meeting their MPs and ministers on their own ground, which helps convey a sense of the atmosphere, pressures and real-life working conditions in politics. It complements visits by MPs to laboratories.

The day enables researchers and MPs to meet informally to share information and form personal relationships. The visiting scientists come from all disciplines and institutions and are very carefully coached beforehand on how to prepare for their meetings, especially on what not to say (e.g. 'Don't mention the 'M' word!'). The accent is on advocating science in general, not individual disciplines or organisations. However, for success, the relationships formed need to be nurtured and followed up, as a single day of lobbying by scientists solves nothing.

MP visiting program

Most scientific organisations welcome MPs into their laboratories, but very few have an organised program of visits designed to reach across the Parliament over time. Often there is a focus on government MPs and a neglect of opposition MPs who may, one day, occupy the government benches. Equally, there is a focus on ministers and senior figures and a neglect of junior backbenchers who will one day become ministers and prime ministers.

The rules for a successful MP visiting program are as outlined above: the information must be clear and attuned to political needs, the topic of funding should be avoided unless raised by the guest, and great efforts should be made to demonstrate the *value* of scientific outcome to the community or nation, as distinct from just the scientific process and its equipment.

MPs, by and large, enjoy visiting scientific sites and ask lots of penetrating questions, which are mainly orientated to how the community is going to react to the new science or technology. From this point on, the relationship can be nurtured and a valuable two-way flow of information and feedback generated.

'Science ambassadors'

These are prominent non-scientists, such as industrialists, financiers, artists, media personalities, religious and community figures who have agreed to advocate for science to government and the wider community.

The aim is to demonstrate to politicians that science has influential community backing, which they can ill afford to ignore. This advocacy is designed to provoke interest, attention, curiosity and even a little apprehension on the part of the politician.

'Ambassadors' can be used to advocate on specific issues, or for the general principles of national scientific investment, literacy and knowledge sharing. It is important to reinforce their arguments by having them also appear in the media, so the politicians 'get it in both ears'. The media is often how politicians assess the political significance of an issue: 'no coverage, no issue'.

Briefing notes

A way for the scientific institution to keep up with the astonishingly fast pace of politics is to create a series of backgrounders or briefing notes on critical issues, which can be easily accessed as the issues surface.

These need to be carefully tailored to MPs' information needs, not just recycled scientific reports or press releases. They need good, clear executive summaries with dot points. They need to make the information available at several layers of complexity and detail. They need up-to-date cost–benefit estimates or risk appraisals. They need to provide 24-hour contact with the relevant experts.

They should be available several ways – as paper handouts, by email, news feeds, on a special website and via the Parliamentary Library. They should be available to senior public servants responsible for briefing ministers and committees, and to the staff of MPs. They can be available to industry or community lobby groups to reinforce their own arguments. They can provide background to political journalists when an issue breaks. All this helps to ensure that consistent scientific information flows into the political system from several directions at once.

Customer value analysis

In politics, customer value analysis (CVA) involves regularly polling and interviewing MPs for their views on the state of science and technology, and their scientific information needs. It is a cheap and effective way to make sure the scientific message is carefully attuned to the political need. It is also a way of helping the scientific institution to appreciate that politicians are actually their customers, not merely a milch cow for funding. The technique is described in detail in Chapter 4.

We recommend two kinds of research: short, simple questionnaires covering a cross-section of politicians and political parties, and more detailed qualitative analysis using a focus group of 'typical' MPs to help interpret the results and provide detailed feedback. One way to do this is to recruit the members of a parliamentary standing committee dealing with science.

Having a staff member work in a politician's or minister's office is another invaluable way of attuning the organisation to political requirements for information, the preferred form and modes of delivery.

Key account management

Key account management means having a central person or office within the scientific institution whose job it is to keep track of all the various contacts that are being made with the world of politics and government to help coordinate them. This person ensures that the approaches are consistent and prevents the organisation from tripping over its own feet or contradicting itself; in other words, to monitor and supervise the total relationship. The concept of key account management comes from the advertising industry, but it has features from which open science can benefit.

One of these is the use of a central computer database that logs all current and past contacts with government so that any individual with access can see at a glance what has gone on before, what is happening now, and who is talking to whom. Apart from better organising the institution's own dealings with government, this also helps it to work far more effectively with other scientific bodies or industry lobbies, and so present a more unified and disciplined front to government.

Two cultural changes are essential for this to work well:

- The 'begging-bowl' model of science lobbying must be replaced with the 'science of service' model.
- The dog-eat-dog academic tradition of pressing one's case for support over that of 'competing' disciplines and institutions must be replaced by a whole-of-science or science-with-industry partnership approach.

Relationship-building with both politicians and bureaucrats requires dedicated and persistent investment of time and people. To begin with it may involve initiating contacts with MPs or senior public servants to offer assistance. As they come to view the organisation as helpful to government and the policy process – as opposed to lobbying in its own vested interest – they will begin to call for advice.

Cost–benefit analysis

A salient difference between the worlds of science and government is that government is largely run by economists who want to know what something costs and what it will return. In the era of economic rationalism, this demand was focused almost exclusively on dollars and cents. Like other industries, science now has to account for its activities according to the triple bottom line of economic, social and environmental outcomes. Added to this, it is having to report against an ever-growing list of bureaucratic performance indicators and targets that make no distinction between the outcomes of science (which are very hard to predict) and the outcomes of welfare or taxation policy (which are more easily modelled).

Yet few scientific programs can produce even a half-credible set of economic figures from their bottom drawer when questioned about the value of their work, let alone the other information politicians need to make a decision.

Scientific institutions suffer from an understandable desire to devote every spare dollar to research activity and as little as possible to 'administration'. There is sometimes a naive attitude that any dollar not devoted to research is a dollar wasted. However, science dollars do not simply materialise, but rather must be argued and cajoled for, negotiated, justified and accounted for. Failure to invest a strategic percentage of an organisation's global income in this activity is liable to lead to a reduction in research funding in the longer term.

As a rule of thumb, every major scientific research program should be accompanied by at least some basic, professional cost–benefit and social and environmental benefit analysis. Ideally, a proportion of all research programs will be subject to independent evaluation by these criteria. Some attempt should also be made to garner community feedback at an early stage in the research, both to help lubricate adoption or commercialisation and to assure government that the money is being spent in a way the community approves and welcomes.

Old habits, like 'science knows best', die hard. But die they must if science is to persuade the 21st century community of its value and relevance, respond more fully to its needs and share its findings more widely and equitably. Those institutions able to adapt most swiftly to the age of modern social and economic accountability are likely to find themselves at a Darwinian advantage over the rest.

INDUSTRY

Private industry is one of the most effective and valuable means for science to share its knowledge with the community, the nation and the world at large. The successful transfer of science and technology to industry is viewed as an important indicator for defining the value of a research establishment to the community it serves and a determinant of the level of public, as well as private, funding it receives. The market orientation of industry to its customers imposes an exacting discipline on science to try to make its knowledge outputs as useful and useable to the wider community as possible.

This ought not to conflict with the duty of a publicly supported research institution to carry out public good research, although there are plainly times when it does – or when the two make uneasy partners. There are also times when tensions arise among researchers and units – between those who see their role as focused on the needs of industry and bringing in commercial funding, and those who regard themselves as operating on behalf of the public good or pure discovery and who see private funding as tending to distort or detract from this ideal. There is no simple solution to this tension – indeed it is healthy – but in communicating the outcomes of research and in achieving open science there is unquestionably a viable middle ground.

One way to harmonise the apparent dissonance between commercial science and public good science is to regard companies with whom an institution works as the proximate, or immediate, customer and society or consumers as the ultimate customer. This dual focus allows the research body not only to meet the needs of its immediate industry customer, but also to ensure that the knowledge product is more likely to be widely adopted by society.

If private companies had a monopoly of wisdom about the market and perfectly understood their clients, they would never go bankrupt or fail. There would be no bears on the stock exchange, only bulls. Yet scientific institutions frequently treat industry partners as if they were omniscient with regard to the needs of the marketplace. The result of this naive trust is that many technologies successfully transferred to industry fail to become fully commercialised, because the company misjudged the market or the cost of reaching it. As a result, the investment in the original research is not fully realised, and the knowledge not optimally disseminated.

We propose that the best way for science to help industry is to have a strong understanding of industry's own customers and *their* needs, right down to the ultimate consumer. If this process is observed rigorously, the public interest – at least as expressed in the views of the public itself – will also be served. Furthermore, by having had input into the early part of the research process, the community will be reassured that its views and values are being taken into account in the development of new products or processes, and will be more receptive to them when they become available.

The objection will be voiced that this violates commercial confidentiality – but there are plenty of ways to incorporate representative community views into a confidential process without letting one's competitors in on the secret: companies use them all the time when doing their own market research and product development.

A good example of this process at work is the 'Cassandra Report' developed by Food Science Australia, a food research institute affiliated with Australia's CSIRO. The first step was for researchers at Food Science Australia, using their scientific and technical prowess, to project all sorts of exciting new products and processes based on state-of-the-art knowledge in their field. The second step was for leading food companies to outline their ideas of future products and processes and the trends they foresaw in the processed food market over the coming decade. The third, and critical, step was to show the combined list of ideas to consumers, who were then invited to choose what they wanted. The result was an interesting selection of novel foods and technologies that were scientifically feasible, commercially attractive and desirable to consumers. Interestingly, the consumers sometimes selected quite different products to those that either the scientists or industry had preferred. A significant point about this process is that the inclusion of consumers in the discussion at an early stage can dilute the commonly held notion that industry and science are insensitive to consumer wishes and needs. The result of such processes will be advances that satisfy everyone's requirements better. Science will become more open.

In developing countries, where the gap between a multinational corporation and the rural poor is extreme, the importance of holding such a conversation is even greater. There will often be profound cultural, religious, ethnic and other belief- and values-driven factors that radically affect whether a new technology or product is accepted or not. Scientific institutions in these countries can play a vital role as the facilitators of this discourse, helping to bridge the gap between the poor and science-based industries. One of the most valuable elements in this feedback process is to give large companies a clear idea of what the poor can and cannot afford, what level of technology they want and of the importance of providing free or extremely low-cost knowledge to earn their confidence.

Any global corporation that is serious about expanding its markets and client base in the 21st century will consider ways to provide the kind of knowledge to developing countries that will enable poor people to help themselves and to lift their own incomes and living standards. Although the corporation's ultimate goal may be to sell more vehicles, computers, entertainment or medications, a truly strategic approach to doing business is to enter a conversation with developing countries about how best to help them meet their needs for better agricultural and environmental know-how, improved education delivery, village-scale manufacturing and processing activity, and low-cost water, energy and health-care systems. Corporations arguing that delivery of these sorts of services are the province of governments have yet to come to understand their role as global citizens, or the billions of potential customers out there to be won by such an approach.

Enlightened companies already see these things as their role and their responsibility, both as world citizens and as astute investors, to help to share knowledge among the poorest of the poor and among those who lack easy access to information they can use. Very often this knowledge is of the most basic kind, concerned with food production, housing, health, education, providing clean water, and so on. It is often unrelated to the main corporate enterprise – a mining company that builds a local school, clinic or water supply, for example – but can nevertheless be delivered by endowing local institutions, through science communication, extension services, NGOs and local community groups, education and technical training. In this way, the partnership between science and commerce comes to have a far deeper public good significance.

The rest of this chapter offers ideas for ways that science can communicate more effectively with industry.

Identifying community needs

It is important for science to have a clear idea of the wishes and needs of the ultimate customer – the consumer or general public – when it enters discussions with industry about its needs.

Economical ways to do this are covered in detail in Chapter 4. They consist primarily of:

- independent quantitative research into public needs and priorities
- a literature search of other publicly available survey findings in the relevant field
- careful analysis of the market segmentation for various products and processes
- qualitative, or focus group, research to determine the factors that lie behind strongly held community wishes or beliefs
- customer value analysis

- inclusion of consumer advocates, community, environmental and health representatives on scientific advisory panels
- face-to-face discussion with advertising consultants and marketers who understand the main drivers in particular markets, or mass psychologists who understand the motives behind community beliefs and values
- detailed discussion with any groups likely to be provoked, angered or hostile to the proposed research (e.g. environmental lobbies, religious groups or minorities) to understand their motivation and needs.

Identifying industry needs

Critical for successful commercialisation of science is for the institution and its scientists to get inside industry's head. The following suggestions have been shown to work well by scientists at the National Science Institute of Indonesia [14] and the CSIRO in Australia:

- face-to-face consultation between senior industry executives and research leaders, followed by meetings between technical staff and front-line researchers
- market research and analysis to identify potential for new products and processes, with attention to market segmentation, targeting and product positioning
- regular industry/science priority workshops to define and adjust goals and priorities
- employment of industry specialist staff and trained business managers by scientific organisations
- creation of industry advisory panels and asking them the *right* questions
- adoption of key account management principles
- visits by researchers to industry to identify problems and explore research opportunities
- visits by industry R&D managers to science centres to observe capability and explore research opportunities
- surveys of industry R&D managers' needs and priorities
- joint feasibility studies and market research into new products and processes
- customer value analysis
- regular participation by researchers in industry conferences and workshops
- science centres taking out membership of industry associations and professional bodies

- joint development of industry research networks, seeking to bring in ideas from as wide a field as possible
- industry/science exchange programs, whereby scientists work in industry and industry technical staff work in science centres
- secondment programs that place scientists in industry for periods of 6 months or longer
- encouraging scientific staff to undertake leadership roles in industry and professional bodies
- wider subscription to industry journals in the science library, including specialist newsletters providing advance intelligence of important developments
- subscriptions to the financial media
- inclusion of industry technical specialists in internal seminar series
- staff news bulletins or notice boards (virtual or actual) containing the latest news and intelligence of industry developments
- awards to honour effective partnerships with industry and successful commercialisation
- development of reward and incentive structures, including profit-sharing and royalty-sharing deals, that allow scientists to benefit from having good relations with industry
- joint seminars and training courses in intellectual property (IP) management
- training scientists in listening skills, negotiation skills and the management of high-technology businesses.

Communicating science capability to industry

Many of the initiatives outlined above will allow the research organisation scope to advertise its capabilities and share its knowledge more effectively with industry, because they are designed as two-way exchanges – both to listen to industry's views and needs, and to explain how science can help meet them.

However, there are also many other tactics the science institution can adopt to raise awareness of its skills, capability and achievements in industry. These include:

- obtaining increased coverage of its work in the financial and specialist industry media (see below)
- participating in industry exhibitions, displays, field days, etc.
- collaborating with industry in new product launches
- linking the scientific brand with industry's leading commercial brands in publicity and media coverage

- holding industry briefings on the latest scientific and technological progress (see below)
- including an 'agreement to publicise' clause in research contracts with industry (see below)
- obtaining recognition from government and community leaders for successful work done in partnership with industry
- helping industry to tell its story to government
- providing industry with an up-to-date directory or e-directory of scientists, their expertise and contact details
- making sure that its latest advances feature on the most-used industry websites
- electronic marketing, in all its various forms (email, web, multimedia, social network sites, etc.)
- preparation of publications (paper and electronic) that are carefully crafted to meet industry's information needs (as distinct from the science body's need to promote itself). These should be concise, written from a business perspective, contain hard financial cost–benefit data, case studies of successful science–industry partnerships, and be layered to allow busy managers to read at the depth that suits them
- sponsorship of industry events, awards and scholarships
- corporate gifts and presentations that reflect the organisation's scientific capabilities and skills
- industry 'ambassador' programs, in which highly respected figures from industry undertake to advocate on behalf of the scientific organisation to their industry
- use of knowledge management 'storytelling', proposed by IBM as an effective way to foster collaborative creation of knowledge using a technique thousands of years old amplified through state-of-the-art media and communication techniques
- collecting and publicising 'satisfied customer' endorsements
- collaborating with industry bodies in schools and public education programs and projects
- providing independent product safety and performance testing, and advice on how to correct problems
- being visibly associated with the setting of national or industry standards of quality, safety, performance, etc.
- participating in the setting of industry codes of practice by providing objective scientific measurement and advice

- providing independent scientific advice to industry lobby groups that are seeking changes to government policy
- acting as an independent umpire or 'honest broker' in public debates and disputes between industry and community groups or non-government organisations
- feeding back to industry any findings from the science institution's public opinion research that point to emerging problems for various industries and their products, where these problems can be overcome by R&D.

Business and industry media

The business and industry media are a vital link between science and industry that are too often neglected by scientific organisations. Many industry managers first become aware of the achievements and services offered by a scientific institution via the business news – often about their competitors.

One of the most effective ways to engage a company's interest is for them to read about their competition stealing a march by adopting the latest technology.

Even if your scientific organisation is planning a series of face-to-face meetings with senior industry figures, these will almost never take place in a perfect information vacuum. The executives will already have an impression of what your organisation does, and of its past achievements and current abilities, culled either from hearsay or the media. Like others, business executives watch TV, listen to the radio, read papers and magazines and admire heroes. Like most of society, they glean their general knowledge of what's going on from these sources. For the same reason a farmer ploughs and fertilises a field before sowing a crop, it is a smart move for science bodies to use the business media to work up the ground of awareness in industry.

The principles for communicating with the business media are similar to those outlined in Chapter 5, but here are a few extra tips:

- Science stories should always be cast in business terms and should be angled around business, not scientific, outcomes. Improvements in profit, production efficiency, product design, customer satisfaction, safety, wholesomeness, environmental sustainability and so on are the sorts of benefits sought from science by industry. The best plan is to get a business journalist to write your story.

- Like the general media, business media are more interested in the outcomes of research than they are in the process. It's important to highlight the practical achievements from a science–business partnership.
- Highlight national economic benefits such as export income, employment, GDP, inflation, consumption and efficiency.
- The business media are increasingly interested in 'triple-bottom-line' outcomes, so emphasise social and environmental benefits along with financial benefits.
- The business media like 'hero' stories about top executives. Use real people to highlight how advanced managements use R&D to get ahead of their competitors.
- The business media also like large feature articles that project possible futures for key industries. Top scientists who have a good grasp of the industry they work with are in a great position to envision the future 10 years or more out, and so stimulate industry interest and debate. Arrange suitable interviews or op/ed (opinion) articles in the business media.
- Business media letters pages are a great place for scientists to stimulate discussion of emerging issues – and implicitly advertise their own capabilities.
- The business audience accepts 'advertorials' (paid space in a publication or program in which a specially written article appears), and this can be an effective (though more expensive) way for a science centre to reach a large business audience by recounting its industry success stories.
- Specialist journals that serve particular industries, such as manufacturing, food processing, IT, fishing, the auto industry, mining, energy, and so on, welcome stories from scientific organisations. They are invariably hungry for well-written articles, happy to give them a good, detailed run, and generally provide the most favourable coverage it is possible for a science body to obtain in *any* media. Once again, the clue is to write the story from an industry, not a research, perspective. If you haven't got a specialist writer to cater to this market, use a freelancer.
- Business media love graphics. Provide graphs, tables and other visual aids to understanding with any media release or story.
- Business readers enjoy gossip as much as the next person. Try to deliver your best science-in-industry stories to the leading business columnists, whose writings are the daily fare of business lunches and watering holes.

- Business and industry publications also offer easier opportunities for regular columns by researchers or institutions than the general media. Like all columns, however, you must be absolutely certain you have something new to say each time – not repeat the same old stuff. Also, avoid promoting your own institution – focus on what science can do for industry overall. Use a skilled journalist to ghost it.

Agreements to publicise

Many good science-in-industry stories slip through the cracks because, by the time the commercial partner is ready to tell the story, the research institution has long thundered off in pursuit of something new. The gap of months or years between science leaving the lab and business releasing its product onto the market often means that:

- the scientific organisation doesn't get credit for the achievement
- the company doesn't reap the market benefit of having its latest product linked to a reputable science institution.

Every day, business journalists are inundated by a tidal wave of company media releases announcing new products or processes. Most of these are not used because the journalist has no way to rapidly assess the truth of the company's statement, and will delete it rather than risk becoming a vehicle for dodgy claims, corporate 'PR' or a covert bid to inflate the share price.

However, if the journalist receives media statements from the company *and* from the science agency *at the same time*, then the science body's reputation for objectivity and integrity will reinforce and validate the company's claims about its new product. This can increase the chances of media coverage markedly. It can also greatly increase the amount of coverage received.

For these reasons, it is important to build into a commercial research contract a mechanism that allows both bodies to gather due credit at the time the product goes public (assuming it does go public).

This is the 'agreement to publicise' clause, which binds the parties to work together on the communication of the research outcome.

There is no reason for this clause to conflict with requirements for commercial confidentiality while the R&D is in train. It specifies that, once the period of secrecy is past and the company desires beneficial publicity, the two will work in partnership to obtain it.

Of course, in cases where the research is deemed permanently confidential, this does not apply. Nor would the agreement to publicise clause cause the release of commercially sensitive details. It is intended only to publicise the *outcome* of the research in terms of its significance to the community, consumers or customers.

Experience indicates that two separate statements – one from the company and one from the science body – work better than a single release incorporating comment from both. For the sake of its reputation for independence, it is better for the science centre to make its own statement and commentary on the product.

'Brand partners'

The use of an agreement to publicise clause introduces an important concept in science communication: brand partnerships.

Both partners in a research collaboration – the scientific organisation and the commercial company – have a public identity or brand that has an intrinsic value. In the case of the company, this is readily measured by turnover, profit, share value or a set of performance indicators. In the case of the research institution, the brand value is much harder to quantify, and consists of the accumulated public reputation from its past scientific achievements. However, this can be very high – as in the case of an 'old' university.

The linking of these two brands around a research outcome holds large advantages for both partners. It is a case in which the whole is greater than the sum of the parts. The commercial brand enhances the science institution's reputation for successfully developing and delivering useful and innovative products that are valued by society. The scientific institution lends to the commercial product its reputation for independence, scientific integrity and research quality.

Ways in which a research organisation and its commercial partners can cooperate as brand partners include:

- placing an agreement-to-publicise clause in the contract
- collaboration in product launches
- combining on media publicity
- working together on innovation case studies that demonstrate public, private and national benefits
- joint appearances at science briefings, parliamentary hearings, and commercial and industry forums
- joint presence in media advertising
- approved use of the science institution's name in advertising and promotion by the commercial partner
- approved citation of the science institution's research findings in product promotions
- statements intended to ease public concerns about secrecy, ethics and ownership of intellectual property (see below).

Public concerns

Public opinion research in both developed and developing countries is showing an increase in public concern about:

- commercial secrecy
- ethics
- ownership and control of research results and intellectual property.

Public trust in scientific institutions may generally be high, but research indicates that it falls significantly as the public becomes aware of the extent they are working with industry, especially with international companies. This can have serious consequences for public funding of research.

This phenomenon has become particularly marked with the public debate about globalisation, the growth of the anti-globalisation protest movement and the resurgence of protectionist sentiments in many communities. Science bodies are increasingly forging partnerships with international companies, rather than local ones, because the former can afford to pay for advanced research. Yet local communities and consumers often feel far less loyalty to the products of a 'foreign' company than they do to a local one. Consequently, they disapprove of 'their' science body using their money to work for a 'foreign' entity.

In every case where a scientific body partners a large foreign or international company, or even a local giant, it needs to think very carefully about the public good issues, how they will appear to the public and how it is going to discuss them with the local community. This is important to avoid perceptions that science is working for global companies at the expense of local interests.

For the brand of a scientific institution to be of optimum value to industry, it is essential to address these public concerns through openness and transparency. Excessive secrecy, or even a failure to communicate, will reinforce public suspicions about the value of the national public investment in R&D, and the motives and ethics of science agencies.

Industry briefings

Industry briefings are an effective way to build dialogue and understanding between science and its proximate customers in industry. The following advice was developed by the National Science Institute of Indonesia:[15]

- A successful industry briefing is not a single-day activity. It needs careful preparation, as well as effort and energy.

- It is important to build a sound relationship with industry well in advance. This means being involved in industry activities, taking membership in industry associations, ensuring good coverage in the industry media, etc. Developing personal contact with senior executives and top figures in industry is critical, as they make the decisions for their companies and can be highly influential over opinion in their industry. They can also open many doors.
- Building and strengthening the confidence of industry in our institution is a prerequisite for conducting a successful industry briefing.
- Industry briefings have three purposes:
  - to gather input on industry's needs and how they see future trends in their field
  - to obtain feedback on your scientific products and services
  - to introduce the latest research achievements relevant to industry customers (where these are not confidential).

  These three functions can be carried out together, but it is advisable to focus on one function at a time as this will bring a more effective result.
- Industry briefings succeed best if they are carefully focused on a selected field of high relevance and interest to industry. Selection is usually based on a combination of current 'hot topics' and the science agency's capability and capacity.
- People in industry are normally very busy and their time is precious. We must present our work in a direct and concise way, and provide business analysis of it. At the briefing there must be no distinction or boundary between scientists and industry people.
- A half-day briefing focusing on just one of the three purposes listed above works best. This allows busy executives the rest of the day to do their job. It also looks business-like.
- Provide all briefing participants with a smart and eye-catching information sheet to support information conveyed verbally during the briefing.
- Invitations should be sent out well ahead. Two weeks is the minimum, but for senior executives far more notice is needed, as their diaries are filled 6–12 months or more in advance.
- It is a good idea to announce forthcoming briefings at the current event so participants can put them in their diaries.
- Notify industry media about a forthcoming briefing well in advance (i.e. at least 2–3 months beforehand), bearing in mind their

deadlines (which may be monthly) and how often they publish. Give their readers enough time to log your event in their diaries.
- The media should also be invited to attend the briefing itself, as they can help bring the message to any industry executives who may have been unable to attend in person, as well as to a wider audience.
- Industry participants from previous meetings can also be invited to nominate or bring along colleagues to the next meeting.
- In preparing material for the briefing, great care must be taken to pitch it at the correct technical level for the audience, and some previous research into their qualifications and experience is highly desirable.
- During the briefing, avoid the inclusion of speakers who know little or nothing about the topic, even if they are senior executives of the science agency. Effort must be made to avoid giving industry the impression that science is bureaucratic and obsessed with hierarchy. If you use a non-specialist to welcome the guests, make sure that person's talk is short and relevant.

Communicating with agriculture and the rural sector

Farmers, foresters and fishers, whether in the developing or developed world, are one of the most important target audiences for open science. Nearly two billion people grow food for a living, and most of the world's farmers are women. They manage 40 per cent of the world's land surface, 70 per cent of its freshwater use, the majority of ocean bioresources and one-third of the global atmosphere.

Even in remote, poor and far-away places, food production affects the lives of people living in wealthy cities who, on the surface, might consider themselves insulated from its ups-and-downs. The success or failure of agriculture in developing and developed countries can:

- mean the difference between peace and war[16]
- increase the probability of refugee crises
- have significant impacts on the global economy, trade, jobs and interest rates
- have large impacts on the global environment and biodiversity, and on the ability of the Earth's resources to sustain the total human population[17]
- have significant effects on health and nutrition in both developing and developed countries

- increase the rate of climate change
- determine whether or not the population grows, stabilises or shrinks.

The failure of agricultural development is a trigger for government failure in many crisis-prone regions. Stable political systems are unachievable when people are starving or fighting over scarce resources.

The ability of the developed world to share knowledge about sustainable ways of producing food, fibre and timber in such a way that it is appropriate to the cultures, peoples and settings in developing countries is, perhaps, the central issue of the human destiny in the resource-scarce 21st century. However, it is also important that the developed world more effectively distributes knowledge of sustainable, low-input systems and the production of healthier food among its own farmers. From either perspective, the sharing of agricultural, forestry and marine knowledge will be a primary determinant of humanity's common future.

There are many well-tested ways to share knowledge with farmers, fishers and foresters, such as:

- government extension services
- private agricultural consultants and advisers
- university outreach activities
- promotion by agricultural technology companies
- rural newspapers, radio and TV
- the internet, digital media, mobile phones, CDs, virtual farming websites, etc.
- farmer groups dedicated to achieving improvements in productivity such as 'harvest clubs' and breed societies
- Landcare, Seacare and other groups dedicated to a more sustainable agriculture or fishing industry
- farming or fishing cooperatives
- agricultural schools
- field days
- rural shows, displays and exhibitions, theatre
- religious groups
- dissemination of fact sheets
- circulation of extension media to local radio and TV stations
- hotline advisory services
- expert columns written for local newspapers.

One of the most important lessons from years of extension research is that primary producers, whether in the developed or the developing world, prefer to get new knowledge and information from another producer, as distinct from a scientist or extension worker. A key to successful transfer and uptake of knowledge in rural communities is the 'early adopter': the adventurous farmer or fisher prepared to take a few risks with their livelihood to try out a new, possibly more productive, method. This person often performs the invaluable role of integrating a new technique into a traditional production system – of making it work properly in the local culture, climate or farming system.

The early adopter is the producer into whose fields all the other producers gaze – a constant object of interest, suspicion, admiration and cynicism in community discussion. The greatest value in knowledge sharing lies not in the early adopter's successes but in understanding their mistakes and learning how to avoid them.

Such people are often used in government or agri-company field days. They are a mainstay of farm productivity and Landcare groups. However, one place they can be used much more effectively is in the rural media – in farming newspapers and, in less literate societies, on radio. By discussing the challenges they faced in adopting a new piece of technology or method, and by telling their story of failure, error and ultimate success, they are one of the outstanding ways to achieve open science in a rural community.

Another technique involves getting primary producers to discuss new techniques in a setting where they are not afraid of disclosing their own ignorance among their peers. An Australian researcher, who was working to reduce agrichemical use in Asian grain farming systems, recounts that by simply videoing farmers at work in their fields and then replaying the video to the village at night – to the delight, amusement and fascination of all present – he managed to stimulate a wide discussion about better ways to farm without chemicals. No single farmer was particularly exceptional in his approach, but among all of them there were valuable take-home lessons and clues that the video brought together as a starting point for the adoption of new methods.

At another level, a scientific institution placed a dozen pages of editorial in every issue of the leading magazine for farmer-innovators. The articles were drafted by scientists and then re-written by agricultural journalists in language more suited to their readers. By this low-cost means, the science agency was able to reach the top 20 per cent of primary producers: the innovators and early adopters who set the pace for the others. A survey revealed that four out of five of these farmers said they had changed their farming methods as a result of advice gleaned from articles in the magazine – an astonishing adoption rate. A parallel case in a developing country was the use of radio interviews with local sugarcane growers describing their experiences in trying to adopt and adapt new farming systems and technologies. Likewise, a marine management agency used video interviews with fishers to raise awareness about sustainable fishing practices. These were distributed to every boat in the fleet, and could be watched as the crews sailed to and from the fishing grounds. The growth in use of the web, multimedia, podcasts, networking sites and other forms of digital technology by primary producers is greatly increasing the power, scope and interactivity of these methods.

Although the science agency can pick and choose among the various methods and technologies for delivering its knowledge, the use of a real farmer, forester or fisher as the messenger is likely to remain a constant ingredient in success.

The rules for preparing a communication plan aimed at rural audiences are:

- segment the audience and understand their differing needs
- make sure the information and messages apply, and can be understood locally
- use spokespeople who are credible to a rural audience (rather than to scientists)
- carefully select key messengers based on credibility and respect
- network with relevant professional, community and interest groups
- set up the process as a dialogue, an exchange of information, rather than a monologue
- use the same terminology and language as the rural people speak
- be prepared to 'get mud on your boots' to find out what local people think and want.

Chapter 7

Communicating with the public

In the 21st century it is the people who have the greatest say over science and whether or not it is adopted.

This was heralded by a groundbreaking report by the UK House of Lords Science and Technology Committee, which concluded that 'direct dialogue with the public should move from being an optional add-on to science-based policy making and to the activities of research organisations and learned institutions, and should become a normal and integral part of the process'.[1]

The report noted a 'crisis of trust' between the public and science, brought on by issues such as mad cow disease, the GM food debate and the pressures for change caused by information technology. It found that the British public had 'much interest, but little trust' in modern science and technology. A number of issues underpin this lack of trust:

- the perceived purpose of the science is crucial to the public response
- people now question all authority, including scientific authority
- people place more trust in science that is considered 'independent'
- government, institutional and commercial secrecy are a major issue
- many scientific issues also embody social, ethical and moral aspects, and excluding these invites hostility
- what the public finds acceptable often fails to correspond with the objective risks seen by science
- underlying public attitudes are people's values. These cannot be challenged or ignored lightly.

These findings apply fairly universally, with similar experiences in European countries, New Zealand, Australia, North America and in parts of Asia, including developing countries. Impelled by the wavering trust of its public, Britain's science policy establishment has been thinking creatively about the issues of science and society, the democratisation of science and the engagement of the public in the research process. It sees 'a new mood for dialogue between the public and science' involving:

- consultations at national level
- consultations at local level
- deliberative polling
- standing consultative panels
- focus groups
- citizens' juries
- consensus conferences
- stakeholder dialogues
- internet dialogues
- foresight programs.

Without such techniques, the risk is that – by taking a science-centric rather than a society-centric position on the knowledge economy – national research directions will look great to science policy supporters, but risk rejection or delay by society.

For more than a century, and in most countries of the world, science and science policy have been driven by those in the field while the rest of society has been largely excluded, or cast in the role of 'cargo cult' recipients of technological largesse. For most people, this experience of exclusion from science – the most fascinating of all fields of human discovery – begins in the schoolroom. The experience then continues through life, resulting in policies designed to promote science and technology or 'the knowledge society' that somehow never quite seem to achieve traction, the main reason being that society has already been disenfranchised. That is when the public starts to erect roadblocks in the paths of researchers and innovators.

The fundamental issue is that the public holds the ultimate sanction over science and its institutions. Through politics, it can shut down almost any line of inquiry, program or even research institution that is perceived to flout community standards or present a risk to society. It can deny its adoption or reject it in the marketplace. It can demand all sorts of legal and regulatory strictures. This has happened to a considerable extent in the field of biotechnology and seems likely to have a significant impact on nanotechnology too. The reason is quite straightforward: failure to consult the public, engage them and earn their trust.

The UK Lords concluded that all the techniques listed above have value, help the decision maker to listen better to public values and concerns, and offer the community some assurance that their views are being taken into account. These increase the chance that new technologies or scientific advice will be adopted, although scientists who have been working on the same issue for most of their career cannot reasonably expect people who have only just learned about it to make up their minds in a few weeks or months.

'They are however isolated events, and no substitute for genuine changes in the cultures and constitutions of key decision-making institutions,' the report continues. 'A meaningful response to the need for more and better dialogue between the public and science in the United Kingdom requires us to go beyond event-based initiatives like consensus conferences or citizens' juries.' The conclusion is striking: *the very terms of reference and procedures of scientific institutions must be changed, to open them up to external influence and input from diverse sources.*

'Nonetheless, in modern democratic conditions, science like any other player in the public arena ignores public attitudes and values at its peril. Our call for increased and integrated dialogue with the public is intended to secure science's 'licence to practise', not to restrict it,' the report added. The Lords then followed up with a firm recommendation that public consultation should become an integral part of doing science – not an optional add-on.

The British Council, in its report on the democratisation of science, lists the following essential preconditions:

- openness
- transparency
- responsibility and accountability
- independent advice and research
- appropriate technological trajectories
- meaningful dialogues
- skills and education policy development
- equality in the distribution of knowledge and technological solutions
- initiatives to forecast, recognise and resolve conflict.[2]

Most scientific institutions, if they were scored honestly against each of these criteria, would receive low marks, even a fail. Indeed, Brian Wynne, noted British science communication academic, argues that the continuing failure of scientific institutions to adapt their institutional culture to an age that favours dialogue is a significant contributor to the growing mistrust felt by society towards them.[3]

The rest of this chapter looks at practical ways to build a dialogue with the public, in addition to those methods already discussed for the media, government and industry. It also sets out some of the steps necessary for this to occur.

THE INSTITUTIONAL CHARTER

Most scientific institutions have a charter, an act of Parliament or some formal instrument that causes them to exist and defines their role and purpose. In the majority of cases, this will refer directly to scientific research or discovery, to being a place of 'learning' (which includes the research function, alongside the educational one) and, on occasion, to a duty to 'publish' their findings and discoveries. The last point is normally interpreted in the narrow sense of publishing in the academic peer-reviewed literature, not necessarily in the wider public domain.

Although these charters prescribe that a representative council or board, consisting of worthy citizens, oversees the institution, very few lay down requirements for wider consultation and discourse with the public. This is an obvious and critical omission. A scientific institution that is not committed to public dialogue by its charter or act risks becoming an anachronism in the 21st century – the century of open science.

The same applies to scientific programs initiated by legislation or regulation. Under democratic principles they, too, should have consultation and public discourse built in.

POLICY BODIES

The obligation to hold dialogue with the community should be a major function and activity of science policy organisations, such as government bodies or academies, instead of an adjunct to what they presently do. Academies in particular are the natural forums for these exchanges to take place. However, with exceptions like the UK's Royal Society and America's Association for the Advancement of Science, academies tend to be underfunded for the task of explaining science to society and engaging in a nationwide discourse.

Best practice

Within the spectrum of scientific institutions, universities and policy bodies resides enormous, but very variable, experience in ways to communicate more effectively with society and its various elements. The pooling and sharing of this experience and the creation of national best-practice guidelines for communication and dialogue is desirable. This will save the majority of scientific institutions from having constantly to 'reinvent the wheel'.

International leadership

There is a vacuum in international leadership in the fields of science communication and public dialogue and consultation. Given the absolute importance of the sharing of knowledge between the haves and have-nots in this century, in order to overcome the 'six crises', much esteem awaits the nation or institution able to demonstrate global best practice, set standards and encourage others to follow suit.

FUNDING AGENCIES

Bodies responsible for funding research, especially from the public domain, have a particular interest in ensuring that the benefits flow efficiently to society. They can help to achieve this not only by adopting a higher level of commitment to openness, consultation and dialogue with the public, but they can also powerfully influence the attitude of researchers and their institutions by making communication and dialogue a mandatory requirement of every grant. Until this rather simple step is taken there will be insufficient stimulus to the scientific community to change its culture from being closed to being consultative and open.

INDUSTRY BODIES AND COUNCILS

Industry bodies and councils also have a strong interest in having a public – or consumers – who are engaged, enthusiastic and responsive to new technologies and processes, rather than mistrustful or hostile. They, too, can send influential messages to the research community that public dialogue leads to more successful research adoption – and less waste of research funds and time. With science now so highly geared to the needs of industry, views such as this from industry will be extremely positive in securing cultural change in science and reduced levels of suspicion and mistrust in the community. This in turn benefits industry.

PROFESSIONAL SCIENCE ASSOCIATIONS

The associations and institutes that represent physicists, chemists, biologists, earth scientists and all the various tribes of researchers are the guardians of professional ethics, standards and practices for their members. Many of them already place a reasonably high priority on communicating, although this is generally from the narrow view of wishing to recruit more bright young people to the profession or to garner greater public recognition and respect. There is an enormous opportunity for professional associations to articulate the importance of their members holding dialogue with the public, and so become a powerful force for changing research cultures. This will be high in the professional code of any enlightened body.

GOVERNMENT AGENCIES

Government bodies responsible for standards, safety and other technical and regulatory matters have a high responsibility for preserving public confidence in science and technology, and in the ability of governments to regulate them successfully. Being composed chiefly of technical people and bureaucrats, they have tended to rely more on the aegis of their authority rather than effective dialogue and communication aimed at greater public understanding and support. This is changing rapidly, with more and more of these agencies acknowledging that public confidence is better obtained through dialogue than an overbearing assertion of technical expertise or by invoking the law. Their communication skills, though not in general high, are improving. A recent setback has been the excessive emphasis on corporate and government stakeholders in framing overall strategy. This approach has the downside of de-emphasising communication with the public, and may leave the agency focused on the needs of, say, big government and big business rather than those of the wider community (who actually pay for the science with their taxes).

COMMUNICATION METHODS

The following communication methods have all been used to engage the public in a dialogue about science and technology issues:

- **National and local consultations** can be set up in which government or scientific organisations create forums for the public and scientists to interact on particular issues of interest and concern. They advertise these, call for submissions and encourage the media to cover them.
- **Citizens' advisory panels** explore particular issues and aspects of science or scientific institutions. There are two main approaches: the first being to use well-known and highly regarded citizens who are not connected with the science, such as prominent lawyers,

philosophers, artists and community leaders who are regarded by the public as having high integrity. The second is for the panel to consist of the nominated representatives of particular citizens groups and NGOs, such as consumer, environmental or animal welfare associations. This version is somewhat more politicised, but has the advantage of bringing critics inside the tent. The panel can either be standing or convened for a single task, although the former is more likely to gain public recognition and trust.

- **Lay members of science committees** can provide particularly valuable advice about how society is liable to receive or react to new developments. Because of their wide contacts across a spectrum of the community, journalists make useful members, as do lawyers, sociologists, psychologists, philosophers, former politicians, science communicators, and consumer and environmental advocates.
- **The internet** provides an interactive way to communicate with the public but, in practice, is generally used as an 'information dump' and this can be counterproductive. Despite its vogue with scientific and government institutions, it suffers the major drawback of being inaccessible to very large groups in the population, including the elderly, lower socio-economic groups, the vision-impaired, the illiterate and all those who simply have no access to computers. It also contains an ocean of absolute garbage whose presence may devalue or compete with serious messages. However, blogs, chat-rooms and social networking sites where the public can gain immediate responses from experts to their questions are nevertheless one of its attractive features, as is its ability to signpost other sites of interest and relevance – including opposing viewpoints. Few scientific organisations use the internet to initiate dialogue with the public, and this remains an oversight.
- **Public opinion research**, both quantitative and qualitative, can provide very effective and up-to-date snapshots of what the public, or segments of it, knows and thinks about science and technology issues. This was discussed in Chapter 4.
- **Media analysis** and **journalists' workshops** are a valuable two-way mechanism for understanding how the public sees and reacts to new technology as they are presented by the media. They also gather the media's own views on it, and their impressions of their own audience's opinions about it.
- **Consensus conferences** comprise a representative group of citizens and selected experts from science, industry and government meet to discuss an issue in depth over several days, and produce a consensus

report or judgement covering all those points on which they can agree and noting where they dissent. Although uncomfortable for science and industry, these discussions help them to understand far more clearly what they are dealing with regarding public attitudes. These conferences can be broadcast and covered by the media, thus enlarging their audience reach into the community, and various forms of public feedback can be built in.

- **Citizens' juries** are similar, but instead of citizens and experts negotiating a consensus position, the citizens simply deliver a verdict. This may, of course, lead to polarisation of views – but can give science timely warning that a particular approach may not be acceptable to the wider public.
- **Foresight projects** are where science and technology experts project various futures arising out of present-day knowledge and technological trends and expose them to feedback from various public, industry or government audiences, and then publish a summary of the views. These are helpful in getting the public, or subsets of it, to focus on future issues and see the need for new technologies.
- **Industry seminars** are an important way for science and industry to come together to plan the best ways of introducing a new technology. They are generally of greater value where they include representative views from the community.
- **Newsletters** and **e-letters** are useful in communicating between organisations and with the media, rather than with the public. The key to success in newsletters is to provide readers with concise material that is exclusive, informative and useful to the audience. Few meet these criteria. Feedback should always be encouraged and published.
- **Labelling** of food and other consumer products is a vital way to convey scientific information about health, safety and environmental aspects of technology. However, it is rapidly becoming so technical that it is meaningful only to few – and this alone is alienating to many people. Every effort should be made to keep it simple and relevant, avoid jargon and encourage feedback. An important step is the use of citizens' advisory panels to help decide what should and shouldn't go on labels, and how it ought to be explained.
- **Radio** and **video** are valuable means of communication in areas where literacy levels may be low, provided the people can receive them. The most effective technique is to present discussion about technology involving consumers or users who are typical of the local community, including points both for and against it. Where possible,

radio talkback can be used to engender discussion in the community. Even though some callers may seem mindlessly critical, they nevertheless give vent to community feelings and frustrations, permitting people to feel their views are at least being registered. Science should not fear talkback, but should be patient and constructive.

- **Open days** and **open laboratories**, where the public can stroll through and observe the scientists at work, are a useful way of demystifying research, especially if there is an opportunity to ask questions and exchange views.
- **Specialist media** such as farming papers, hobby magazines, professional and industry journals and specialty websites are ideal for in-depth articles exploring new technologies and discussing the pros and cons. Public comment and debate can be sought on the internet.
- **Shopping centre displays** are useful, especially for consumer-related science issues, if there is a knowledgeable communicator, or a good website, to answer questions. There must also be a means for the public to record its views and feelings.
- **Museums, science centres** and **galleries** present excellent opportunities to engage the public – as they have done for more than 150 years. Modern museum philosophy calls for far greater interactivity, 'hands on' experiences and direct engagement with the public than the 'glass case' mode of earlier times. More and more 'exhibits' are going online, and are thus able to reach and engage a wider audience. Original use is being made of social networking and other technologies to display and debate ideas and invite audiences to contribute their own thoughts and exhibits. Physical exhibits are often designed as 'roadshows' capable of being transported to many different venues. Ideally, these also feature public seminars where open discussion of controversial science can take place. For 'visitors', dialogue about science is now part of the 'museum experience', not merely the passive reception of information; for this reason museums can and should be used much more fruitfully by R&D organisations to gather public feedback on controversial new science and technology.
- **Science circuses** and **drama** are an animated and enjoyable way to present scientific concepts to children and young students; care must be taken to allow time for questioning and interaction.
- **Teacher conferences.** There is great value in seeing that all teachers – especially teachers of sociology, social history, general studies, English, and so on – are better informed and updated about modern science and technology, and have a chance to explore the issues

around it, rather than being a source of potentially negative attitudes in both the common room and classroom. This can be achieved through teachers' conferences and in-career training. In primary school, scientific topics can be used to teach basic literacy and numeracy as well as providing an easy and interesting introduction to scientific thinking.

- **Politicians.** Chapter 6 referred to the use of politicians as messengers and feedback-providers for science. Few people are so acutely attuned to nuances in community opinion as politicians, and the well-briefed MP can be extremely valuable for building public discourse around science and technology.
- **Religious institutions** are highly engaged in the community's moral and ethical values, as well as issues such as equity, health and safety. They can play an extremely valuable role in facilitating dialogue between science and the community, sharing knowledge and meaning. They can particularly assist science by transmitting community values into the research thinking and helping to ensure new technologies are used in ethical ways or do not disadvantage particular groups in the community unduly.
- **Non-government organisations** sometimes have a ideological barrow to push but can nevertheless be an important vehicle for science to tune into the articulate concerned in the community – environmental opinion, for example – and should be included in the process of discussion by scientists, not shunned. It will also surprise (and hurt) scientists to learn that the more famous advocacy groups enjoy a higher credibility with the community than do many scientific institutions. This is due to a perception that they are on the side of the public, whereas science is perceived by many to be sliding into the corner of private vested interest.
- **TV chat shows, opinion columns.** All of these form the basis of the discourse that goes on in a democracy, and very often important people, such as politicians and the media, pay attention to what is being said. Though they are seldom used in science communication, they are nevertheless an important forum and outlet for scientific ideas and discoveries. It should be regarded as the duty of scientific institutions to insert a scientific view into this form of social discourse. They are useful for gauging public reaction to controversial new technologies. They have the added advantage of a large and disparate audience.

SCIENTIFIC PUBLICATION

Most science is withheld from the public until it has been (a) peer reviewed and (b) published in a credible scientific journal. Although there is good sense in this, there is growing questioning of whether important scientific findings, funded with the public's taxes, should be withheld from the public at the mere behest of the editor of a commercial journal. Scientific organisations need to consider the harm they may do to themselves if they are seen to have withheld important information – and it is wise to explore the pros and cons of announcing important findings before the usual process is complete.

Linked to this is the commercial desire of scientific publishers to withhold information from society until they have printed it, in order to make money from its publication. This form of trade protectionism is gradually being overcome with the arrival of online open access journals, but much publicly funded science is still held back for the pleasure and profit of publishers. Smaller journals are learning that early release of findings can in fact benefit them by drawing the attention of a wider audience to important articles they will carry.

The expansion of online publication and its immediacy is changing the landscape of scientific publication dramatically, introducing new ways of sharing knowledge in a timelier and publicly accessible fashion. This is the way of the future, and will soon be seen to benefit both science and society more so than the 19th century journal model that has prevailed until now. At present, vast tranches of the world's knowledge lie inaccessible in paper libraries – inaccessible in that society cannot easily access them and they are written in a way that is often unintelligible to the lay reader, or else they form but a small fragment of a wider body of knowledge.

In the 21st century, the premier role for science communication will be the translation of scientific findings in journal articles into plain language and advice of immediate use to society or industry. To some degree this is already happening – but on a scale not nearly large enough to address the 'six crises' referred to in the first chapter. What should be automatic for virtually all science is still largely treated as optional.

A STATUTORY SCIENCE COMMUNICATION BODY?

Countries sometimes debate the need for a national science awareness or communication entity to oversee and help to improve the standard of dialogue between the research world and the rest of the community.

In our view, a statutory body is only desirable to the extent that it obliges the scientific and science policy world to take the issue of science communication as seriously as it does the issues of discovery and invention, and to recognise that it needs to lend it equal weight and emphasis in policy.

A science communication agency could help by providing best-practice models to scientific institutions to harmonise the culture of communicativeness in science – but this job can be done just as readily by existing entities such as government science departments or academies. The last thing the need for good communication should cause is the formation of fresh layers of bureaucracy.

Finally, it is not by any means clear that the public would trust a government-owned science communication body – and might even regard it as a professional spin-doctor for science. Nor is it clear that individual science institutes and associations would cooperate fully with it.

Overall, a national science communication advisory council seems a sounder way to go, providing it has influential and active connections to the key scientific institutions, universities, funding agencies, academies and policy bodies, and is not dominated by academics but consists substantially of communication professionals and representative citizens. Even then, its role will be limited to advice, guidance and best practice – to influence, rather than to regulate or enforce.

After all, the best people to communicate science to the public are the scientists themselves, with all the assistance and support they can receive from their institutions, their professional science communicators, their governments and industry partners, and especially from the public itself.

Scientists have the priceless qualities of enthusiasm and love of their subject, a little of which goes a long way in the communication game.

Chapter 8

Talking to the world

Swedish futurologist Åke Andersson has described the emergence of a society in which minds are webbed together along great axes of intellectual development and high technology, reaching from city to city at the speed of light, transcending national borders and spanning the globe. Along these axes thoughts, ideas, discoveries, collaboration and creativity flow freely as the world's best minds network together in real time. Beyond the axes, however, lie great hinterlands of darkness: places were knowledge, enlightenment and advanced technologies do not reach, or penetrate only in a fitful fashion.

With the exponential growth in the internet in the past quarter century, even into developing countries and some remote regions, Andersson's vision has come to reality. By 2010, some two billion people around the planet were sharing thoughts, ideas and information at light speed.[1]

Something even more profound may be happening to us at the species level. Early in the gestation of the human embryo, the higher brain begins to form. Individual brain cells link into filaments and begin to transmit messages back and forth. As the network ramifies throughout brain and body, at some indefinable moment the capacity for thought is born. The embryo becomes a person capable of sensation, feeling, reason, dream and imagination.

Worldwide, at this moment, a 'planetary' brain is also in the forming as individual humans – the neurones in this giant mind – connect in real time with one another and with vast networks of fellow beings. A capacity for 'global thought' is being born, which was almost inconceivable in the age of printed literature. However desirable or undesirable this may appear to the individual, it is hard to deny its reality, or the capabilities – good and bad – that it will unleash. Among the good are the ability to share the knowledge essential to overcome the 'six crises' that confront humanity, rapidly and with most of humanity.

Advances in information and communication technologies create the possibility for a more universal sharing of science, technology and the resulting knowledge than at any time in history. However, the world of science does not yet seem geared for this. The moral issue of the 21st century is whether the flood of new knowledge serves to benefit the privileged few – or the majority; whether the 'brain' exists to serve the whole body, or simply itself.

Scientific research institutions have never previously enjoyed such an opportunity to share their discoveries, achievements and findings with so many people all around the world, so quickly, easily and at such a low cost.

The most affordable techniques involve the internet and other world media, which are fast becoming ubiquitous. Even in regions not yet 'wired' for internet services, satellite television, radio, mobile telephony, papers and magazines penetrate widely. Furthermore, knowledge is increasingly reaching remote places through education, extension services and commerce. The challenge for scientific institutions is to select the most effective and low-cost among these many avenues to disseminate their knowledge – and to use them to build a system for sharing and discussing new knowledge with society as it emerges.

THE INTERNET

The internet offers the most powerful and pervasive means to communicate new scientific discoveries and advances to the most people. Its most cogent elements are:

- the power to aggregate news and information from trustworthy sources and present them in a coherent and networked way
- the fact that people use it to select the 'news they can use' or the information they prefer.

This makes the internet an unforgiving medium for those who simply use it to advertise themselves or brag about their achievements – but a friendly one for those who supply what the audience is seeking.

A number of science websites specialise in reporting the latest scientific news to their audiences. These are regularly visited by science journalists, students, industry research managers and others interested in remaining in touch with the advances in knowledge and with disseminating discoveries and advances still more widely.

Some of these sites (such as Australia's ScienceAlert) publish science stories from reputable research institutions free of charge, while others (such as America's Eurekalert) levy a fee or subscription to post an announcement on their website.[2] A third category (such as ABC Science) operate as internet news services and do their own reports based on material received. As a rule, we suggest using all the sites that are free, and paying for publication only on those others that can demonstrate an influential audience in the target group(s) sought by the institution. Here, mere numbers of daily 'hits' are not enough to demonstrate impact: ask to see evidence of the audience, where it comes from, its linger time on the site, and so on.

As the generations change, the internet is gradually supplanting the print media, especially newspapers, as the preferred source of news, information and entertainment. One of the impacts of the global recession was for internet usage to overtake newspaper readership in the US for the first time, and to close the gap in most other Western economies. Young people in particular prefer using the internet to reading papers or magazines. There are clear signs that the internet is replacing TV as the main information source for the general public and, as new services emerge and media fuse, this trend is likely to accelerate.

The internet is therefore a vital way for a scientific organisation to reach both the public at large – including the affluent global public – and the media who will in turn disseminate the story to other users. Through the use of blogs and social networking sites such as MySpace, Facebook, Twitter and YouTube, the internet has become a valuable tool for generating public interest in science and obtaining uninhibited feedback about it – albeit from a very poorly characterised sample of humanity. Indeed, many scientists now use these sites to share information and vision about science itself, with the added bonus that the interested public can join their discussion.

The ability of people to 'choose their own news' is one of the internet's most important features. Finding novel ways to insert science into these self-selected feeds and websites is one of the most interesting challenges facing science communicators, and one that is generating many original approaches though, as yet, few clear winners.

GLOBAL MEDIA

The global media remains the most powerful tool for the transfer and sharing of human knowledge. In advanced societies, almost all adults gain everything they know about new science and technology from the media. In many developing countries, too, the media is the primary fount of new ideas. Although educators might not see it quite that way, the media is already providing societies with a form of lifelong learning. Its golden quality, from a communication perspective, is that it allows discussion and debate to flow in both directions. The media is pleased to carry scientific ideas to the community, but equally concerned to carry the community's reaction and opinions back to the scientists. The result is a better fit between science and society and more rapid take-up of the best new ideas.

For a scientific institution to access the global media is not difficult, but does require expertise and the commitment of staff and time. For this reason, few attempt it. Yet, having a global reputation is one thing that can help assure the longevity of a research organisation and its funding in the global century.

There are good reasons to develop an international media presence:

- to share knowledge more effectively, with more of humanity
- to alert other scientists around the world to the work of the institution, leading to useful partnerships and collaborations
- to attract investment in research from both public and private sources, and to greatly enlarge access to global venture capital
- to build the international profile of the organisation as a means of attracting the best and the brightest research staff
- to build the reputation of the organisation as a means of achieving more widespread adoption of its research outcomes
- to enhance its own world outlook and shape its research to global needs
- to contribute productively to global thought, debate and policy development on pressing issues.

Given these advantages, it is rather remarkable how few scientific bodies look beyond their local and national opportunities, both for awareness and for investment. It is equally remarkable how often those who are seeking to build an international profile do so by employing commercial image-mongers, rather than by communicating their genuine worth and their real achievements to humanity.

LEADING MEDIA

The most effective method for communicating the value of a scientific research organisation to the global community is to deliver factual accounts of its discoveries and achievements to specialist science and technology journalists from the world's leading media, both print and electronic.

This is not so difficult a task as might be imagined, as most countries have one or two media – usually quality national newspapers or national broadcasters – that dictate the news trend for most of the other media in their country, and are closely followed by them and by decision makers.[3]

All that is necessary is to deliver well-crafted science stories to the several hundred leading papers or broadcasters, ascertain the names and email addresses of their key science, health or environment correspondents, and then seek their permission to deliver appropriate science stories to them. Most of these journalists are keen to receive good quality science stories from other countries in order to keep abreast of global developments in their field. What they re-publish will depend on the news value of the story in their country and at the time of writing. However, the returns from such coverage can be impressive in the form of new investment in research, new partners and customers, and wider international recognition and stature.

An up-to-date email list of the world's leading scientific correspondents is a pearl beyond price for the scientific organisation with a global perspective and international ambitions. Needless to add, regular work must be put in to ensure the list remains current. Professional science media delivery services such as SciNews offer this specialised service.[4] Many international news services can also provide delivery of general media stories.

EMAIL AND RSS NOTIFICATION

To avoid spamming science journalists (and others) with long stories that clog their email systems, a good alternative is to use an email notification service or RSS (really simple syndication) feed. These alert them of the existence of a new story, with just a heading and a 'teaser' sentence from it, linked to the fuller story on a website. If the topic interests them, they simply click through to it.

Some skill in crafting the headline and teaser is necessary, so that enough solid information is provided, but enough is unsaid to 'tease' the reader's interest. This allows journalists to select what they read and to work more efficiently, without overloading them with information irrelevant to their interest. They can subscribe or unsubscribe at will.

INTERNATIONAL NEWS AGENCIES AND NETWORKS

An effective way to reach leading national and global media, as well as others, is through international news services such as United Press International (UPI), Associated Press (AP), Reuters, Australian Associated Press (AAP), Agence France Presse, AGI (Italy), Novosti (Russia), China News Service, Interfax, ANTARA (Indonesia), BERNAMA (Malaysia), Deutsche Presse, Panafrican News Agency, BBC News, FOX News, CBS, CNN, and so on.

A news story delivered to these agencies will be distributed widely to international, national and local media, as well as the business media, and will often be reproduced as a news agency report on the world news or business pages when their in-house science writer is otherwise occupied.

Some agencies offer a media release delivery service in which they charge for the distribution of releases to other media outlets, but in our experience the cost of this is high, the precision of delivery to a science audience low and the timeliness inexact, compared with the science institution maintaining its own distribution system or using a specialist service such as SciNews.

ELECTRONIC NEWSPAPERS

An increasingly important target for distribution of science news are the electronic editions of famous newspapers and broadcast news services published on the web, such as the UK *Telegraph*, *Times* and *Guardian*, the *Washington Post*, *LA Times* and *New York Times*, *Frankfurter Allgemeine*, *The Australian*, *Le Monde*, *Pravda*, *La Repubblica*, *Jerusalem Post*, *The Star*, *Straits Times*, *O Globo*, *New Zealand Herald*, *Jakarta Post*, *Times of India*, *People's Daily*, and the BBC, ABC, CNN, and so on.

Although, technically, the electronic newspapers reflect the content of their newsprint big brothers, their editors are quickly discovering that the electronic audience is not the same as the paper audience. It has different standards of technological literacy, for example, is more adept at searching for and locating information in the electronic medium, and has a higher interest in technical and scientific issues. This means, in practice, that many electronic editors are seeking to subtly distinguish their product from the paper edition and are more receptive to stories about science and technology. Electronic 'papers' also publish continually, being refreshed many times in a 24-hour period, whereas their print sibling normally has only one or two editions. They therefore have a far greater hunger for news stories.

It is worthwhile for a scientific institution with a good story to send copies of the announcement separately to both the parent newspaper (science writer and news desk) and to the news desk of the electronic edition. It may appear in one but not the other, and it is good practice to hedge one's bets, as they do not necessarily share their copy as well as they should. Most electronic newspapers provide an email address for delivery of media statements on their websites.

CDS, DVDS AND ELECTRONIC CARDS

These are all useful ways to convey detailed information in multimedia format, especially video, long reports and complex graphics. However, they depend on the recipient having the time and patience to browse the contents and are, in any case, now largely being overtaken by the internet as download speeds increase globally. As for other media, the rule is to make sure you understand your customers' information requirements and reading/viewing habits before you invest a lot of time and money in trendy electronic publications that feel good internally but are not widely used externally.

Plastic business cards with readable digital memory that provide basic information about an organisation, its achievements, capabilities, services and contact details are quite useful. The card can be hooked to the internet to provide greater depth of information to the user. An advantage is that the information can be tailored to the needs of the individual whose name is on the card, or the particular client.

CABLE TV

Cable and satellite TV services with a focus on news, science, discovery, business and current affairs appear to have an almost insatiable appetite for well-made TV science stories to stock their magazine programs and news bulletins, and to use as fillers.

Their requirements vary, but most in demand are items of 3–5 minutes length, produced in international TV and digital formats and with the capacity for the voiceover to be dubbed in different languages. It is important to emulate TV news style with brisk, stimulating plain-language reporting that carefully avoids the slightest suggestion of 'PR', but instead adopts an objective journalistic tone and is accompanied by compelling vision.

Having gargantuan appetites for new material, many cable TV companies will broadcast material that meets their production standards free of charge. A great advantage of using cable TV is repetition – the same item may be screened several times, at different times of the night and day, over weeks or even months.

Some cable companies will seek to charge scientific institutions to broadcast their material, which can be expensive. It is advisable to do careful research into the viewing audience profile in order to determine if this is a good investment.

THE ORGANISATIONAL WEBSITE

The following are some ground rules for using the website to share knowledge more effectively.

- Great science websites are based on sound communication principles and planning. 'Look' comes second to content and ease of navigation.
- *Don't* promote the organisation; promote the scientific achievements and outcomes.
- Avoid bright colours, flashy layouts and gimmicks that detract from scientific credibility.
- Use images, but keep them to thumbnails or low resolution pictures. *Avoid* large and high resolution pictures, complex multi-image graphics, animated devices, music, movies that take up too much bandwidth for easy download by the public (who often use older, slower computers and, if rural, may still be on dial-up internet).
- Refresh the site continually: constant variety, novelty and change are the secret of generating traffic. Stale sites lose their audience and reduce 'linger-time'.
- Link up with as many related sites globally as you can: the stronger your network, the greater the traffic is likely to be and the higher you will rate on the search engines.
- Ensure on-site search engines are capable of using not only scientific but also lay terminology, so the public can use them easily. Include a search by industry or profession function.
- Provide plain-language definitions of scientific terms used on the website, as scientists sometimes employ the same word differently in different disciplines.
- Give careful thought to the meta-data and other search criteria that will attract people using search engines to your website. The choice of key words can be critical. Ideally they should be terms in common usage that describe the scientific work of the organisation (unless the intent is to restrict visitors to specialists).
- Always provide contact points for further information.
- Provide easy access for different categories of visitor, behind which they can find specialised help and contacts (e.g. special sections of the site catering for industry, students, the public and research partners).
- Avoid a 'blank front door' or index-style home page that forces visitors to click through several screens before getting to the information they want. Put your best achievements and services on the home page and vary them often.
- Offer a notification service to inform visitors by email, RSS feed or social network site when new items are posted on the site, thus encouraging them to visit regularly.

- Classify scientific information in lay terms (e.g. industry sector or societal issue) rather than by scientific discipline.
- Provide 'layered' information, with plain-language summaries on top followed by more specialist information for particular audiences (e.g. industry) and finally, hard science.
- Provide a 'dial-an-expert' database to help industry and the public locate the appropriate scientific expert within your institution quickly and without multiple calls.
- Provide lots of signposts to partners, industry clients, government and non-government organisation websites. Be helpful to visitors seeking a broad range of information on a topic and you will become a regular internet 'crossroads' for them.
- Provide an email/phone inquiry handling service to refer queries to the right expert or area of research if they are unsure what advice they need.
- Offer blogs, chat-rooms, visitors' books and other opportunities for public comment, feedback and questioning.
- Use the website to poll various audiences for opinion and feedback. This can be done either actively (i.e. solicited responses) or passively.

FREE FEATURE SERVICES

A promising technique, in an age of cost-conscious media, is to provide a website containing professionally written feature articles on your science, offered free of charge. A surprisingly large number of these are picked up and run by regional, national and international media, many of whom face very tight budgets and cannot afford to pay for much freelance material. It should be noted that, in principle, for an organisation to provide a pre-written article conflicts with journalistic ethics. However, if great care is taken to ensure that the article is objective, unbiased and written by a professional freelance journalist with an established name and track record, this difficulty can be avoided. However, this means that the scientific organisation must adopt a 'hands off' stance with regard to the content and style of the article, which is the province of the journalist who writes it and the editor who accepts it for publication.

The key tenets of a 'free feature' service are:

- articles should be written by professional freelance journalists, and be scrupulously objective, not organisational PR
- they should be accompanied by quality colour images in a range of sizes or resolutions, and including photos, graphics and vision

- the site should be refreshed with new features regularly to encourage editors to visit it
- it should notify editors by email or RSS whenever new articles are put up.

'OP/EDS'

The media is often grateful to receive opinion articles written by (or ghost-written for) eminent scientists contributing to public debate on topical issues. These are a good way for the scientific institution to signal its engagement in important and current issues.

It is desirable to have a journalist ghost-write or edit the scientist's opinion piece, because few scientists are masters of the exacting art of opinion writing. The essential element is brevity – most 'op/eds' are between 700 and 1000 words in length, and often have a strict maximum. Structure and style are critical, and traditional scientific writing is not suitable.

A good op/ed opens with a strong expression of personal opinion, not a recital of historical facts. It is crisp, punchy and provocative. Its aim is to stir readers into response, for or against, rather than recite a lot of data or educate the reader. It does not seek to be particularly balanced – but rather to advocate for a particular view or policy.

It is an excellent way for a scientific body to 'test the water' of popular opinion if it is working in a controversial field. When targeted at global media, it is also a way to project the scientific institution's capabilities and identity to a world audience.

DIPLOMATS AND EMBASSIES

Almost all embassies, and many consulates, have an official whose job is to monitor and report on scientific and technological developments in the country to which they are posted. They will usually be designated as the scientific or trade attaché.

These officials provide an effective route for conveying news of scientific discoveries and advances to foreign governments, who will frequently disseminate it to their own scientific institutions, industry and potential partners and to various community groups. The result can be greatly increased opportunities for international scientific collaboration, overseas investment, commercialisation and adoption. It makes sense to include these diplomatic officials in the distribution of science news announcements and reports, and to include all local embassies on your fax/email circulation.

Another way that diplomats can assist the process of sharing knowledge is to channel science news to your own country's overseas embassies and diplomats. This provides your overseas representatives with up-to-date news and information they can share with their contacts in foreign universities, industry, research agencies and government. The best method for notifying this information is by email or RSS feed.

ALUMNI ASSOCIATIONS

As universities become more international – offering courses offshore and attracting foreign students to their home campuses – there is scope to use the overseas alumni network as a way of sharing knowledge, reaching new science partners and investors.

Alumni networks often contain people who, as a result of their overseas education, have become influential in business or government in their home country, yet who still cherish an affection for their university, scientific agency or the country where they studied.

Alumni can be kept up to date with the latest scientific advances by email, RSS and the internet. However, it is a good idea to develop a specialist online newsletter to maintain their engagement.

PROFESSIONAL ASSOCIATIONS

Many professional communicator and journalist networks can help disseminate a science story internationally. They include bodies such as the international conference for the Public Communication of Science and Technology (PCST), science writers and broadcasters' organisations in many countries, SciDevNet (which serves the science-for-development community), the Research and Media Network, the Profnet university information officers association and International Association of Business Communicators (IABC).

INTERNATIONAL AMBASSADORS

As mentioned earlier, an effective technique for drawing global attention to important scientific issues and developments is the use of world-famous figures or 'ambassadors'. These should preferably come from a field or profession unrelated to science so that the focus of news attention is on why they have chosen to associate themselves with this particular issue: creating the curiosity factor.

An ambassador could be a well-known former head-of-state, a famous sports hero or media identity, a musician or rock star, an eminent academic or religious leader – any person who can bring attention and credibility to the issue of concern.

A successful example of this technique was the *Future Harvest* campaign developed by the Consultative Group on International Agricultural Research (CGIAR) to promote global awareness of the importance of international agricultural science with the aim of reversing a decline in investment resulting from apathy about the world food situation.

The campaign recognised at the outset that most people outside the profession of agricultural science find it of limited interest, and do not readily perceive its importance to their own lives – especially if they live in big cities and in developed countries. The goal was to build a wider community of support and understanding for research into sustainable food production in the developing world.

The campaign was built around five 'pillars', each directly affecting the lives of every person on the planet, every voter, every politician:

- **Food for peace** demonstrated the link between agricultural failure, misgovernment and conflict.
- **Food for growth** demonstrated the link between agricultural success, economic growth, trade and stable government.
- **Food for the Earth** promoted sustainable agriculture as the solution to many of the Earth's large-scale environmental problems.
- **Food for health** promoted agriculture as the basis of improved nutrition to overcome the most common forms of death worldwide.
- **Food for people** promoted the idea that prosperity can help bring about lower birth rates.

Among its 'ambassadors' were former US President Jimmy Carter, South African Archbishop Desmond Tutu, Nobel laureates Peter Doherty, Oscare Arias and Norman Borlaug, movie star Jane Fonda, Grameen Bank developer Muhammad Yunus, Queen Noor of Jordan, Francine Cousteau and rock group Hootie and the Blowfish.

The messages they delivered to the world media and both global and international decision makers were based on the five pillars, and argued credibly by organisations such as the International Peace Research Institute, Oslo, and the World Conservation Union (IUCN).

The argument was backed up by the scientists working at the world's international agricultural research centres, whose research achievements and views demonstrated the importance of investing in agricultural research. *Future Harvest* is a model public awareness campaign in that it achieved global attention at low cost through well-conceived, credible messages, and highly respected messengers.

Such an approach can be adopted for almost any important field of science.

Chapter 9

Communicating new technologies

The first time the public caught sight of the motor car, they didn't much care for what they saw at all: a noisy, blasphemous object that transported people without visible means of propulsion at the appalling speed of six or seven miles to the hour, frightening the horses and leaving a trail of squashed poultry in its wake. In Britain they insisted that a man with a red flag walk before it to warn all innocent road users of its imminence, and to restrain the monster to a sedate four miles per hour until society had had a good, hard look at it. It was a perfectly reasonable reaction in the circumstances, because the motor car has since gone on to kill millions.

Humanity has responded with caution to the advent of a new technology many times. Innocent innovations such as pasteurisation and margarine were originally protested on the grounds that they represented attempts to vitiate the food supply and starve the lower orders. The Luddites took a famous dislike to early factory equipment (which subsequent industrial history vindicated). Australians had an initial distaste for food irradiation, nuclear energy and recycled water, which they still haven't entirely got over.

Technical innovation has been going on for at least 2.3 billion years, since the first tool makers in Ethiopia began to shape complex tools from flakes of quartz. By about 1.7 million years ago, pre-humans had figured out how to use fire to make their meat more digestible and to protect themselves; by 750 000 years ago, they were going on short ocean cruises between islands; and, by 650 000 BP, they were building the first true houses in caves out of timber and animal hides, complete with fireplaces. All this took place at least half a million years before *Homo sapiens sapiens* darkened the evolutionary doorstep. One of the things that our ancestors became very adept at was avoiding premature death. The social mechanism for identifying, anticipating and avoiding danger has played a central role in our emergence as a species.

Scientists and innovators sometimes complain that humans are too risk averse, too conservative and too reluctant to try new things. This overlooks our million year history of startling innovation. However, these scientists also forget that innate caution has also served us well for several million years. We are remarkably good at adopting and adapting to novelty, but at the same time we have learned to be extremely careful. We know from immemorial experience that new things are not always good things, and even good things bring bad things with them. The canon of popular fiction fulfils the public belief that, even though the intentions of the researcher may be honest, tinkering with the natural order may lead to calamity. Dr Faustus, Dr Frankenstein, Dr Jekyll, Dr Strangelove and other caricature 'mad scientists' are the yardstick by which their real-life counterparts are frequently (mis)judged. And, some might add, with climate change we have been hoist with our own petard.

Scientists too often bewail the media's apparent obsession with 'bad news'. The media covers bad news such as accidents and disasters mainly because it recognises an insatiable market demand from its readers and audiences for them, and because the media knows that an unrelieved diet of 'good news' will only provoke scepticism among its consumers: life just isn't like that, society tells them. The obsession with bad news and danger is what keeps humans on their toes, their survival instincts sharp, and their scientists productively employed (figuring out ways to make the world a healthier and a safer place and cleaning up the deadly consequences of previous great technological advances).

In communicating about new science and technology, it is essential to engage both facets of the human character – the innately adventurous and the innately cautious.

Recent efforts to deal with this go by the jargon term 'risk communication'. However, because so many technologists, bureaucrats and industrialists are prone to interpret this as 'making the public believe that the risks aren't really as great as they fear' (a completely misguided and essentially dishonest view), it is preferable simply to stick with the term communication, defined as a two-sided conversation and an exchange of knowledge. Interestingly, the public's definition of 'risk' nowadays is often 'We weren't consulted' – and, from their point of view, the less they were consulted, the riskier the issue appears to them and the more the people trying to foist it on them must have to hide. Thus, trying to reassure the community by explaining about comparative risks – 'The technology is a thousand times safer than flying in an aeroplane or crossing the road' – is unlikely to solve the problem of mistrust: the opposite in fact. It is now increasingly understood that society sets its own pace for the adoption of new technologies and innovations and prefers to be consulted. Although this pace cannot easily be hastened artificially, it can be retarded dramatically by trying to force-feed the community with something suspect or unpopular – or just by not asking their views. On the whole, public engagement works better for building trust than one-sided assurances.

A reason scientists do not always enjoy absolute trust from the community is their tendency to put the most optimistic interpretation on their work. If you believed all the public utterances of scientific institutions you would be convinced that science only disgorges an unrelieved fount of blessings – nothing bad will ever happen as a result. This clearly defies real life experience and provokes scepticism. Of course, we all like the world to think well of ourselves and of our work, but practical experience has taught society that most technological 'improvements' have down- as well as upsides. It may come as a blow to researchers to realise that the public is as sceptical about scientists in some respects as they are about politicians: the general view is that the proof is in the pudding, not in the promise.

Chemists are occasionally heard to protest that Rachel Carson's doom-laden predictions of chemical catastrophe in *The Silent Spring* have proven greatly exaggerated, and if the world just had a little more chemical knowledge it would appreciate what exquisite things chemicals really are. This, however, is a view that ignores human experience over millions of years since we first learned to avoid the red berries because they were deadly and the serpent for its fangs. Then we learned to dose Socrates on hemlock, to despise Lucretia Borgia and Dr Crippen for using chemicals to do away with people, and heavy industry for its misdeeds at Minamata, Serveso and Bhopal. Society is understandably restrained in its enthusiasm for poisons and poisoners. Chemists don't always appreciate this.

However, this is one of the main reasons why the modern community is so intensely cautious, suspicious and untrusting when it finds new things in its food supply, untoward effects in its medicines or environment, foreign-owned companies taking control of local industry and local scientists working for them instead of for the local public.

The traditional response of the scientific community has been to attribute these reservations to the community's failure to understand science, and to try to fix things through public understanding of science campaigns and the like. However, recent research suggests that the primary assumption behind such activities – that ignorance correlates with mistrust of science – may be flawed. In an article on public response to GM foods in Europe and in the US, Gaskell *et al.*[1] found evidence that higher public knowledge of science in Europe correlated with greater suspicion and caution, while poorer knowledge in the US was accompanied by greater acceptance of GM food: the more they knew about it, the more cautious they were. Tendering evidence to the UK House of Lords, Sir Robert May, the former UK Chief Scientific Adviser, presented survey data suggesting that people in some European countries have a better understanding of scientific method than people in others, and that people in those countries display less unmitigated enthusiasm for science. This, said Sir Robert, is 'exactly as it should be, because the more you understand, the more you understand that things are complicated and advance makes for change, which produces unintended consequences'.[2]

Although more research is needed, it is tempting to theorise that a more scientifically literate community is likely to be much more demanding and cautious than a scientifically illiterate one. Thus, as science literacy rises, it is essential to engage the community more deeply in discussion about the findings and applications of science.

However, when one considers all the advances that have been adopted in the last 100 years, it can be seen society is not as averse to new ideas as some scientists imagine. Humanity continues to advance at a cracking pace and a little caution about life-changing and 'disruptive' technologies is perfectly rational and appropriate.

Superimposed on this, however, is the overwhelming pressure of technological advances across the board, and the counter-reaction taking place in society. In absolute terms, humanity is passing through its most innovative period ever: people have adopted and coped with more innovations in the past four generations than in the previous thousand. A study of human development soon unearths the indigestible term 'punctuated equilibrium', which simply means that human technological development appears to have gone by fits and starts – periods of rapid advance followed by long periods of stability or relatively subdued progress. For example:

- it took about 1.2 million years for the earliest stone tools to evolve into the Acheulian culture
- it was a further 900 000 years before Mousterian technology replaced them
- this lasted 70 000 years or so before being eclipsed by modern stone technologies around 30 000 years ago
- these lasted only 25 000 years before bronze came on the scene
- bronze reigned for only 2500 years before iron appeared, and so on.

Each step is progressively shorter. However, the fact that there *are* steps is not easy to deny. Humans like to develop a successful new technology and then use it for a while until they feel the need for something better. The surviving hunter–gatherer societies of the Earth are the perfect example of cultures that developed all that was needful for survival in their environment, and settled down to enjoy it – it is the influx of alien technologies and cultural beliefs that has been the major force in their destruction. For humanity overall, a situation in which new knowledge emerges almost every day, requiring new skills and new adaptations, is beyond all our previous experience. It is very stressful.

It is therefore no surprise that parts of society appear to be suffering 'innovation fatigue' – a tendency to call for go-slows and moratoria when confronted with some major new development – and there is a rising resentment and suspicion towards the transnational scientific–industrial complex that is the fountainhead of so many new products and technologies. The habit of many high-tech companies of pouring out new versions and updates compounds this unease among citizens already annoyed, frightened and resentful at the pace of change. This 'fear of change' is superimposed on our natural caution where new things are concerned.

A third, as yet poorly defined, factor is the growing resistance towards globalisation. Although, on the surface, this might seem to be 19th century trade protectionism in fresh guise, in reality it is quite different: most national governments are signed-on when it comes to globalisation, but growing bands of their citizens, including many younger ones, are not. Their concerns are with things such as cultural imperialism, the obliteration of local industries and traditions, the brutality and selfishness of global capital, blind technocracy, the rate at which industry is devouring the environment, and so on. Regardless of one's political views on these matters, it is important to note that the interests of a large worldwide industrial concern do not always coincide with those of a local community, and this mismatch causes continuing friction. Although at present rather inarticulate when compared with the big socialist movements of a century ago, the political potency of the anti-globalisation movement ought not to be underestimated by those working in the scientific and technological industries – not least because it uses the latest communication technology to mount its global protests, with powerful effect.

In the early 21st century, science and technology thus find themselves confronting a society in which many people appear to resist the advent of new things – a situation vastly different from the post-World War II era of infatuation with new technologies. Behind the resistance, as we have noted, lie eons of human caution and a rising suspicion about global industry. These developments should sound warning bells in every research laboratory about the importance of developing more effective dialogue with society.

As long ago as 1989, the US National Research Council released a study in which it advocated 'an interactive process of exchange of information and opinion among individuals, groups and institutions,' which it termed risk communication. For the reason given above – that is it subject to misinterpretation by science and technology practitioners – we prefer not to use this phrase, and consider that communication, a 'two-way activity based on trust, respect and openness', is better as a basis for bringing society into contact with a new technology.[3]

Here are a few principles for doing this.[4]

Responsibility

It should first be recognised that the responsibility for ensuring effective communication rests with those introducing the new technology, not with society attempting to find out what it isn't being told. Communication is an intrinsic part of science, and science that seeks to change society without communicating is behaving irresponsibly towards the society that supports it.

Understanding the public

As outlined elsewhere in this book, an understanding of the public, its views, values and needs is critical. This can be obtained by 'reading the public mind', quantitative and qualitative research, media analysis and other techniques, such as the use of consensus conferences. It must always be borne in mind that there is no single 'public' but a large number of segments, any of which can become a bitter opponent if ignored or mishandled.

Credibility

Information will only be trusted by the public if it is seen to come from a trustworthy organisation without hidden agendas or compromising interests and connections. Even then, the information it provides must be seen to be fair, transparent and responsible. It must also be in plain language.

Balanced information

The goal of science communication is *not* to make the public accept the new technology, but rather to provide accurate, balanced and useful information on which it can make up its own mind. This includes admitting possible risks and downsides as well as potential benefits. It recognises scientific doubts. Many scientific organisations and their managements are reluctant to do this and need to be reminded of the importance of honesty and transparency.

Expertise

Information given to the public must come from the best research sources available and may cover a wide range of disciplines, many of them outside science. These may include fields such as ethics, theology, the law, social science, environmental science, commerce, politics and the humanities.

External critics

Information provided should not only acknowledge scientific differences of opinion over a new technology but also the views of external critics. If it fails to do so, it will appear one-sided and polemical – and less trustworthy. Science should never be frightened of public criticism, but instead welcome it as a chance to put the evidence on the table. The information provided should give priority to the immediate concerns of consumers and citizens.

Clarity

Lack of clarity and use of jargon in presenting a message can annoy and offend the audience, and fail to inform them. It is advisable to test messages before releasing them, to avoid misinterpretation or affront.

Accessibility

Communication must take place in a wide range of media that are readily accessible to the general public, and over a long enough period of time for them to satisfy their information needs and register their views. Because the goal is two-way communication, every method must provide some way for the public to respond and must invite its opinions. The more interactive the method, the better.

This may all seem like rather a tall order for a simple scientific organisation. However, another way to view it is that the introduction of new technology is in reality a community-wide partnership involving researchers, government officials, industry, the public, interest groups, the media, the education system, and so on. New technology is the outcome of a pan-societal conversation. It will reward the scientific institution that initiates it.

Having started the conversation there are a number of essential steps for it to be carried forward.

GOVERNMENT

Governments are finding themselves left behind by the mad rush of technology, trapped between the conflicting pressures within society for faster progress and greater caution. Politicians are urgently signalling that they need forewarning of new technologies so they can consider what legislative and regulatory measures should be taken (although, as previously pointed out, it is not always easy to focus politicians on the future when their obsession is the immediate present).

Government agencies – forged in a bureaucratic tradition that the public can be politely frozen off the doorstep and told to mind its own business – find the challenge of the information society mystifying – especially when critics can splash their opinions all over the world before the first interdepartmental committee has even sipped its tea. They, too, are starting to see the importance of having better listening skills and younger, more congenial spokespeople to explain what they are doing. As for science, the old 'trust us – we know what's best for you' refrain no longer works for bureaucracy and, like open science, 'open government' is more than a passing phase or catchcry.

Public acceptance of a new technology often depends on the confidence the public has that government and its agencies are on their side and protecting their interests. Scandals such as mad cow disease, GM food, medical mishaps and chemical pollution have caused many in society to wonder aloud whose side the bureaucrats are really on. As a consequence, government endorsement of a new technology is no longer as reassuring to the public as it once may have been.

Thus, a high priority for a scientific institution with a new technology is to brief all levels of government about it thoroughly, to help government understand not just the technology but also its wider implications, and to give an honest account of possible downsides or societal objections. The aim is to shore up public confidence in the ability of government to manage and supervise the introduction of new technologies. Ways of doing this include:

- commissioning reports to government from independent scientists or institutions
- briefing key parliamentary committees, ministers and interested MPs
- face-to-face briefing of senior bureaucrats
- workshops with government officials responsible for the field
- joint media awareness activity around the laws and regulations that safeguard the public interest
- developing joint consumer information activity
- collaborating on issues such as national guidelines for the release of new technologies, and the development of regulatory frameworks.

COMMUNICATION METHODS

The main communication methods are outlined in Chapter 7 on page 132.

THE GM SYNDROME

During the 1990s and early 2000s, a wave of hostile public sentiment emerged in both developed and developing countries towards GM foods. This was more pronounced in some countries than in others, but was dramatic enough for some international investment houses to advise their clients to sell off their biotech shares and for a great many transgenic research programs to be terminated. The experience of GM food may be taken as a precedent for other disruptive technologies launched too hastily on an unprepared world – nanotechnology being a case in point.

Although subsequently fanned by the media, concerned and hostile public sentiment was clearly evident in Australian public opinion research years before the first 'Frankenfood' headlines broke in newspapers. In other words, science could have seen it coming had it chosen to inquire and to listen.

Dissecting the various strands of concern that arose in the course of the debate:

- The community did not understand why food had to be genetically modified in the first place.
- The community was angry that large corporations and scientists were attempting to dictate what the public ate, thus infringing one of the basic human freedoms.
- The community had fears about food safety issues resulting from gene modification.
- There was a failure to adequately inform the public, regulators and others about the approach of the technology, how it worked, and what it meant.
- GM food was introduced to hundreds of millions of humans without advice, consultation, permission or any evidence of proper safety testing.
- The industry initially refused to label GM foods.
- Early gene modifications were seen to benefit multinational companies and some farmers – but not consumers.
- The technology caused ethical and religious dilemmas posed by cross-kingdom transfers (e.g. a flounder gene in a tomato), leading to perceptions that the technology was 'against Nature' or 'playing God'.
- There were environmental concerns about the consequences of transferred genes shifting to other species, or interacting with existing genes in their new host or in the ecosystem in unpredictable ways. To the public, this raised spectres of other experiments 'gone wrong', such as thalidomide, cane toads and DDT.

- The new technologies were dominated by a handful of large US and European corporations seeking to monopolise the global food market (epitomised in the use of so-called 'terminator genes').
- Certain companies attempted to patent all genes discovered, whether human or plant.
- There was anger over 'biopiracy' or raiding of the native gene pool, especially in developing countries.
- There were fears of cultural and biological as well as economic 'colonisation' by big corporations, and the lack of global business laws to control them.
- There was a perception in society that their scientists and regulators were working hand-in-glove with foreign companies to introduce these new foods, regardless of the community's views and objections. This caused loss of trust in the integrity of science and the regulatory system.

The purpose of the list above is not to further berate the poor old biotech industry, which knows well enough that it got it wrong. Rather, it is to ask, with the benefit of hindsight, how these same errors can be avoided in future.

Take the issue of safety as an example. Although it is true that safety testing of genetically manipulated products is now becoming more common, science and industry nevertheless did release foods to tens of millions of people on the assumption that, because a gene was harmless to humans in one life form, *ergo* it was harmless in another. There is no safe scientific basis for this, and it remains an assumption. Indeed, evidence from Australia emerged that a fairly innocent gene – *IL4* – inserted into a low-pathogenicity mousepox virus made it 100 per cent lethal to mice. Danish scientists also showed that insecticidal genes inserted into oilseed rape quickly spread into a weed of the same family. Both experiments justify the public's natural caution about insufficiently tested science. The issue of 'nanopollution' – the ability of nano-sized particles to penetrate human, animals and plant cells with adverse consequences and to disperse irretrievably into the environment – is another issue already starting to loom large. Indeed, many scientists say such things ought not to be released without prior safety testing, although hundreds of nano-products are already on the market and it is already too late.

There are many ways to conduct a public discourse between science and society that will lead to quicker and wider acceptance of new technologies – as well as to technologies that are safer, more sustainable and ethical.

The benefit to science of such a discourse is that it avoids the colossal waste of good research, scientific time and skills, which occurs whenever a new technology is rejected or stalled – simply because no one thought to consult society about it.

While the glare of public attention has been largely on biotechnology, similar concerns are apparent in a great many other disciplines including nanotechnology, quantum computing, robotics, data mining, nuclear energy, xenotransplantation, stem cell science, geosequestration, global climate engineering and various potentially polluting industrial technologies.

A final point on the communication of new technologies concerns the ability of humanity to survive the 'six crises' of the 21st century. The latest evidence indicates science alone cannot resolve these crises – which require profound changes in human behaviour and values to occur as well as the adoption of far-sighted and sustainable technologies. From climate change to ocean and atmospheric pollution, from deforestation to land degradation, from species loss to the growing scarcity of water, soil, energy and other key resources, humanity is beset by challenges of such magnitude and complexity that no 'quick fix' will work.

The technologies of the 21st century have profound social implications because they involve refashioning communities, industries, values and even individuals. These are not 'bolt on' technologies like cars, computers or vaccines that emerge from the factory fully fashioned for use. They require understanding, approval and adoption by almost every citizen in their daily lives. They require changes in mindset, morality, tradition, skills, education and regulation. This cannot be achieved by science pursuing its traditional model of 'look what we've invented for you', but only by a strong, dynamic and open interchange of ideas and views: by a discourse rich and full, in which both sides acknowledge one another as equal partners in both the science itself and in the sharing and application of knowledge.

Chapter 10

Working towards open science

Few university science courses include a mandatory unit on open science. This remains a serious oversight – as science that is poorly communicated is often of little value to society.

Undergraduate science courses, as a rule, are highly effective at instilling the importance of peer review and scientific publication in specialist journals as the profession's main communication practice, but tend to leave it there. As the student progresses to higher degrees, it is extremely rare for much to be done to correct this, with the exception of a few bolt-on postgraduate communication courses for those interested. As a rule, the young scientist will attain full-fledged professional rank without ever having had more than a glancing contact with the obligations, principles and practices of science communication. It is left largely to their native talent and personal disposition whether they then go on to become an outstanding exponent of open science or a poor one.

Even those fortunate enough also to be teachers and lecturers are rarely taught the skills for dealing with the multitude of audiences beyond the campus. Nevertheless, some scientists become brilliant communicators – such as Medawar, Sagan, Attenborough, Hawking, Dawkins, Cousteau, Gould, Flannery and Diamond – who have made the straightforward transmission of scientific complexity and adventure a beautiful and inspiring art. Many other researchers develop into engaging advocates for their field in spite of, rather than because of, their academic background.

The fact that scientists have been criticising themselves and one another for failure to communicate well with society for most of the 20th century and into the 21st suggests that something is wrong. A problem so clearly identified and commented upon so widely should not be beyond such a body of intelligent men and women to solve; yet it has proven singularly intractable.

One reason for this may be found in the not-so-charming British acronym PUS, which, believe it or not, stands for public understanding of science. PUS appears to have erupted out of scientific anxiety: 'Why don't they understand us? Why won't they listen to what we tell them? Why won't they adopt my science?' The problem is that this is a self-centred wish. Its exponents desire society to be just like them: to have a basic grasp of physics, chemistry, maths or biology, to be rational rather than emotional, to examine the data and so on. In other words, they reject the law of human biodiversity. Public understanding of science, at least in its original manifestation, is a misguided attempt to get the public to think more like scientists and less like the public – a largely unattainable aim.

A second set of reasons why the challenge of communicating science well remains unsolved are the things that scientific organisations do, both to fellow scientists who are good communicators, and also to professional science communicators. In the first case, there is frequently persecution, criticism, bullying, spiteful remarks, discouragement and outright gagging inflicted on scientists who wish to communicate and who are good at it. To defy this peer persecution takes great courage, independence, confidence and resolve on the part of the communicative scientist. This suggests there is something amiss with the sociology and culture of science that needs to be set to rights in the early phases of training – the undergraduate and postgraduate years. Later, there is also something gravely wrong with the system of recognition, reward and promotion, in which scientific organisations value published scientific papers but do not value articles published in the mass media that may open the science to millions.

The late Peter Cullen, one of Australia's most communicative scientists, warned: '(scientific) publishing is necessary, but no longer sufficient for survival in science … The media is a key tool in communicating our various messages and positioning our "product". Our product is knowledge. We must build support for developing knowledge, which we find a more understandable term than research. We must also influence the community to change behaviours.'

Explaining this in the context of water science, he added:

We will not achieve our vision of improving the condition of our waters without convincing the community it is desirable and showing them how to do it. Communication is therefore a core function of our organisation, and this is reflected in staffing and budgets. It is also reflected in the amount of time our Board spends looking at our communication work, which is reported at each meeting. We plan our communication in a strategic sense, just as we do our research work, and we demand the same level of professionalism and excellence in our communication activities.[1]

As to the general treatment of science communicators by research institutes, there is also something astray with a system that draws an artificial distinction between those who discover knowledge professionally and those who communicate it professionally. Both functions are equally necessary to society. Communicators are often treated as inferior 'administrators' rather than equal professionals in scientific institutions; while this attitude persists, open science will remain a second-order priority. Looking down on their communicators also has the effect of making a scientific organisation less sensitive to views and developments coming from the outside world, and less able to respond to them.

Peter Cullen said that emphasising professional communication in his research organisation had many benefits: staff seeing communication as core business not just as a research add-on, strategic allocation of resources, larger communication budgets, making media work a pleasure for scientists rather than a chore, and sensitising staff to opportunities to get their message to wider audiences.

In an analysis of the impediments to scientists communicating through the media, Gascoigne and Metcalf concluded that the main reasons were that scientists regarded it as optional, saw it as neutral or negative to their careers, believed that management was unsupportive, and felt they lacked the skills or access to professional advice from trained communicators.[2]

The problem lies far less with the scientists than with the scientific organisations and universities that employ them. Research carried out among Australian scientists by Suzette Searle indicates that more would communicate their specialised knowledge with the general public if it was a higher priority for their employers and a formal part of their job:

For the 1,521 self-selected Australian scientists who participated in the 2007 survey 'Australia's scientists and science communication', communicating their specialised knowledge with the general public was personally important to the majority (74%). Most (89%) agreed that scientists had a responsibility to communicate research results in the public interest and that there was a need for more effective two-way communication with the general public.

Communicating with the general public for most scientists was not formally part of their job. Just over one-quarter (27%) of the scientists surveyed said it was a part of their job description/duty statement or project requirement, although more (38%) were expected by their employer to communicate on particular issues or projects.

Just over half (55%) of these scientists, however, were hindered in their communication with the general public and, of these 839 scientists, just over a third (37%) described hindrances in their workplace such as approval requirements and processes, and an additional 23% specifically mentioned time – the lack of it, to organise, prepare and present their communication with the general public.

Of those who did make time (86%) within 12 months prior to the study, most communicated their specialised knowledge though informal discussions, answering questions from the public through their job, providing information though the Internet and speaking with students and teachers at schools, colleges and elsewhere.

Scientists gave many different examples of the personal and professional benefits they gained from communicating with the general public. These included positive feelings about themselves, their communication and their work, such as satisfaction, enjoyment and self-confidence; and their work and personal success, such as direct participation by the public or co-operation in research, networking and building relationships.

When asked what would help them most to communicate with the general public, the most frequent responses were time for communication followed by opportunities to communicate/gain experience, and training to communicate science with the general public and the media.

It appears that more of Australia's scientists would communicate their specialised knowledge with the general public if it was a higher priority for their employers and formally part of their job.[3]

The problem of a 'closed science', as opposed to an open one, can be addressed in various ways, such as:

- instilling in future scientists, while they are undergraduates, that there is a public duty to communicate with the wider world as well as their peers, along with some basic clues about how to do it

- giving them confidence-enhancing and skill-enhancing training and experience throughout their research career
- rewarding, honouring and acknowledging good science communication by scientists and others
- fostering a culture of scientific openness, listening and sensitivity to society as well as to industry and immediate research customers
- management adopting a strong pro-communication policy and leading by example
- identifying and removing the many unspoken disincentives to communicate that exist in all organisations
- committing to a policy of public transparency and open science
- ensuring the availability of sufficient professionals and resources to help the organisation and individual scientific teams to communicate more effectively
- building communication activity into the delivery of every significant research project, and its budget
- giving communication professionals status, rewards and incentives that recognise the value of their contribution to the open science process.

The remainder of this chapter offers suggestions for ways scientists can become better external communicators and exponents of open science.

Why communicate?

The following reasons for communicating are those put forward by scientists themselves, in Australia, New Zealand and Indonesia:

- to transfer to society the benefits of research
- to help make the world a safer, more prosperous and sustainable place
- to advise leaders and policy makers about the latest progress and its meaning for society
- to obtain opinions, needs and feedback from society about research and its outcomes
- to prepare the public for the advent of new technologies and change
- to help governments make better policy
- to alert industry, other researchers, developers, educators, the media and research users of recent progress to speed the delivery of new knowledge and technology
- to get closer to industry and better understand its needs
- to enhance economic growth and sustainability
- to attract greater investment to science and technology

- to attract young people to a career in science
- to engage those who will pursue other careers but still be users of science and technology
- to share the joy of knowledge.

Why scientists don't ...

Scientists also offer the following explanations why communication fails to happen as well as it should:

- they are too busy
- they mistrust or misunderstand the media and its motives
- they lack confidence in their own skills
- they have not been trained to communicate with external audiences such as media
- they fear the reaction of their colleagues
- they see no career advantage or reward
- they fear their work may be misinterpreted or get into the wrong hands
- they are restrained by confidentiality arrangements
- they fear public reaction to their work, especially if based on a misunderstanding
- they fear criticism from lobby groups and vested interests
- they are restricted by an exclusivity clause imposed by their intended professional journal of publication
- their leaders and managers discourage communication and do not provide sufficient resources to do it properly.

HINTS FOR SCIENTISTS

Here are some helpful ideas for scientists who have received a media request for an interview. They are based on the advice tendered in Chapter 5.[4]

Before accepting a media interview, it is a good idea for the scientist to discuss it with their manager, their communicator, or a colleague with some media experience. This will help to organise the main messages, develop some good quotes and avoid sensitive issues.

1. First establish what the story is about and what your role in it is likely to be.
2. Plan your interview with the same care you would a presentation to colleagues.
3. Make your message simple, brief and non-technical.

4. Write down your two or three main points. Write them several times, in different ways.
5. Find out who the audience is and craft your message for them. Answer their unspoken question: 'What's in it for me?'
6. Develop at least one pithy, memorable quote the media is likely to use.
7. Help journalists to 'get it right' by providing a short, plain-language printed summary of your work, with contact details.
8. Explain the relevance of your work to ordinary people. If possible, quantify (dollars, benefits, lives saved, jobs, exports).
9. Don't say 'no comment'. If you can't comment, explain why.
10. Don't tell a journalist how to write their story or insist on 'clearing' it. (They know their editor better than you do.)
11. Offer to check your quotes for accuracy – but don't insist. The media has pressing deadlines, sometimes only minutes away.
12. If you do have a chance to review the story, don't change the journalist's handling of it – only points of technical accuracy.
13. Avoid using too many figures, qualifications and technical terms that may confuse or mislead your audience. The media will only use part of what you tell them – the clearest part.
14. If you want good coverage, make sure your story is news. If it isn't news, provide an irresistible quote and a strong picture angle.
15. For TV and print media, offer picture or vision opportunities (fieldwork, equipment, graphics, kids, animals).
16. To help flesh out the story provide the media with other contacts: authoritative technical, government or commercial sources of comment.
17. Be accessible – the media will value it and use you more often.
18. Try to be helpful – if you can't answer a question, suggest someone who might.
19. Be friendly and open. Avoid appearing arrogant or patronising – it invites attack. Treat the journalist as a partner in reaching a wider audience.
20. Do not lie or mislead – the media has a long memory. If you mislead unintentionally, correct it as quickly as possible.
21. Don't play favourites among journalists (too much).
22. Make it absolutely clear if you are speaking to a journalist off-record or under embargo.
23. Take journalists and the media seriously – they influence public and government opinion and hence funding and public sanction.

24. Learn from your experiences. Seek experienced feedback. Media interviews are not easy and being a good performer comes only with repeated practice and learning from what didn't work.

STAYING 'ON MESSAGE'

By their nature, scientists suffer an almost irresistible urge to 'answer the question'. However, in dealing with the media this can be a big mistake! To ensure that the media reports what you, the scientist, have to say accurately, you may need to return to your key message several times in the course of an interview. Use this technique to avoid being led by the journalist down a line of questioning you are not equipped, permitted or able to answer.

1. Think carefully about who your ultimate audience is: who are you really addressing via the media? Focus on them and their needs.
2. Remember, it's your interview – not the journalist's. You call the shots. You decide which questions you wish to answer, and how.
3. Write your two or three main points on a card. Memorise them carefully.
4. Find different ways to make the same point: the media will select the best 'quote' or 'grab'. Repetition is good – but vary it.
5. If asked to comment outside your expertise or authority, apologise politely and refer the journalist to the expert.
6. Respond to aggression with courtesy. Smile.
7. If you find yourself asked a question outside your field, or in an area you do not wish to go, return to your main point by saying:
   - 'That's a very interesting point, but the real issue is … (your point)' or
   - 'I'm afraid I'm not an expert on that, but what I can tell you is … (your point)' or
   - 'I can refer you to someone who can explain that better than I …'
8. If you cannot comment, explain why. Then return to your main point. ('I'm sorry, but we have not yet analysed the data on that. Please come back to me later.' Or 'I'm sorry but that is subject to commercial confidentiality at this stage. What I can tell you is … ')
9. With practice, you can control an interview by being fascinating – so that the journalist is side-tracked from *their* issue. Have a colourful anecdote or interesting fact up your sleeve with which to engage them.
10. If the issue is controversial, practise. Rehearse with a colleague or friendly journalist how you can get back 'on message'.

IF THERE IS A MEDIA ERROR

If a media story contains an error, *don't over-react.* It may cause more far more harm than the original mistake. Institutional or managerial outrage is a very poor guide to what to do and should be discounted as it is not externally objective. There are ways to fix the problem that do not involve going on the attack and can in fact lead to benefits both for accuracy and the science. For what to do in the event of a media misreport, see Chapter 5.

PUBLIC REPORTS AND MEDIA RELEASES

The requirements for an effective science media release are discussed in Chapter 5.

PUBLIC SPEECHES AND PRESENTATIONS

Effective presentations are important to the standing of the research organisation in the wider community. They can be a powerful way to:

- inform and educate
- persuade or motivate
- initiate a policy discussion
- report new research achievements
- generate media attention
- attract investment and research partners.

1. Analyse your audience. Work out what they want from you and make sure you deliver it.
2. Identify the two to three key messages you want them to remember and take away with them.
3. Select data and graphics that convey your messages – discard data that wander from the point you wish to make or go into too much detail.
4. Signpost your presentation: 'Tell them what you're going to say. Tell them. Tell them what you just said.'
5. Use clear, simple slides, overheads or PowerPoints.
6. Don't use too many! (12 is ideal for a 20 minute talk.)
7. Keep them concise: 3–6 points per slide and 4–8 words per point. Avoid cluttered slides.
8. Don't repeat the slide points word-for-word. Use different words or give a fuller explanation. Make the audience focus on what you're saying after they've read the slide.

9. Use images, graphics and cartoons for a brighter or more amusing presentation.
10. Avoid messy graphs, lots of lines, tiny print and hand-drawn overheads.
11. Rehearsal is the key to a smooth performance. Practise your voice tone, gestures and body language in front of a mirror. Plan your slide changes and integrate them with the talk. Get someone to watch and criticise.
12. Physical 'props' make the presentation more memorable – scientific samples, things in bottles or test-tubes, live experiments or items that can be passed around the audience.
13. Carefully check the venue beforehand: lectern, mike, acoustics, projector/s, computer, software, pointer, water and anything else you require.
14. If possible, have your presentation pre-loaded and checked. Always carry it with you on a memory stick or thumb drive.
15. Afterwards: review your performance. Seek helpful criticism from people in the audience.

Chapter 11

Sensible rules for open science

Modern science exists and carries out its research with the sanction, and the money, of society. Institutions that forget or ignore this simple fact place themselves at risk, as the sanction can easily be limited or withdrawn.

Because the public, government and industry base their impressions of the value of a research institution on what is publicly known about its achievements (as distinct from its scientifically published work known only to the cognoscenti), it is sensible for it to have some basic rules for opening its science to the wider world – be it to government, industry or the community at large – so its research staff clearly understand the extent of their liberty to communicate. Indeed, it is important to have such guidelines simply to make staff aware they actually have permission to share their science and for them to feel empowered to do so. The work by Searle and others indicates that most scientists feel constrained in communicating by the attitude, implied or overt, of their institution and its management.[1]

Where the institution derives significant funding from the taxpayer, there is a clear obligation to share with society the fruits of the research. If this is neglected, there is a danger the institution will be seen as exploiting the public for the benefit of itself and its special partners, and this will cause negative perceptions to arise. The chief motives for communicating externally are to speed the delivery of benefits to society and to demonstrate that the institution is a sound public investment.

It is remarkable how often the formal charter or legislation of modern scientific bodies, while emphasising the research mandate, either overlooks or downplays the duty to share knowledge with the wider society. However, there is an ethical view that says: 'If the people have paid for it, the knowledge produced belongs to the people, not to the institution, the researcher or research partner.' Any research institution wishing to flourish in the more democratic atmosphere of the 21st century is advised to revisit its charter and amend this omission.

Of equal importance for the institution is to send the correct signals to its research staff about their duty to communicate: that it expects them to do so proficiently; that good research choices depend on the needs of the society; and that good research planning includes plans to share the outcomes. The internal rules of public comment and open science should be framed to encourage, rather than obstruct, this process.

This is particularly important in the light of a tendency in recent years in both Australia and the United States for governments and scientific institutions to try to gag their scientists. This gag has been applied over a range of politically sensitive issues: notably climate change, pollution and public health concerns, fisheries and forestry science, environmental science and other hotly debated fields. These attempts to prevent the public, who fund the science, from learning the scientific facts of a situation not only tend towards the corruption of science but also lead to ill-informed policy that is likely to fail. The gags can be both overt and covert, explicit and implied. Very often their effect is to chill the public discussion of a scientific issue, to attempt to stifle media discussion and to cause timid or politically compliant scientific managements to impose self-censorship on their institutions.

There are many ways to go about establishing sound communication guidelines. The following suggestions are adapted from those prepared for one of Australia's leading research organisations.[2] They make it clear to the scientific staff that they are expected and encouraged to communicate. They open with a statement of purpose:

> *We are committed to excellence in science. We are also committed to delivering outcomes based on our science and to communicating our science.*
>
> *Communication is a part of our charter. It is encouraged by the Organisation and it is seen as essential for the successful adoption of the outcomes of our research by industry, government and other stakeholders. It is vital to the national debate about our common future.*
>
> *The Organisation's claim to be a leading provider of excellent science will only be convincing if we continue to provide the nation with examples of what we are achieving.*

The Organisation only exists because of community sanction. We must never lose that, and one way we can make sure of this is to continue to report what we discover and achieve for the benefit of the community. We must also listen carefully to the views and values of the community about what it expects of our science.

The standing of the Organisation in the wider community gives us scope to influence the nation in favour of scientifically literate policies. It also induces far-sighted and competitive industries to use the results of our research. Our reputation rests significantly on public awareness and approval of our scientific achievements.

Staff are encouraged to communicate with industry, government, the public and media, effectively and responsibly. These guidelines are designed to help you. The Staff Code of Conduct and the Commercial Practice Manual also contain policy statements on communication issues.[3]

What is public comment?

Public comment includes public speaking engagements, submissions to public inquiries, comments on radio, television, the internet and in the newspapers, views expressed in letters to the press, in books, journals, in brochures, magazines or on the internet – in other words, wherever it is likely that publication of comment will flow to the community at large or a significant part of it.

A guiding principle is that public comment should take account of the need for constructive relations with stakeholders in industry, government, the community and within the organisation, and not injure our scientific reputation or standing in the community.

Public comment on scientific issues

Staff have a responsibility to communicate with the public and industry about scientific aspects of their work. The organisation encourages this, subject to various laws and policies. Effective public communication is included in staff evaluation and promotion.[4]

The organisation encourages its staff to contribute to the public debate on scientific issues within their area of expertise. Such comment should always be tempered by judgement and tact.

It is not expected that staff will comment on scientific matters outside their expertise or on non-scientific matters (e.g. politics or religion) unless they are authorised to do so, or they are quite clearly commenting in a private capacity and the name of the organisation is not linked to their remarks.

Departmental heads are formally accountable for judgements exercised on matters of public comment within their department.

Media announcements

Official media announcements are made through the organisation's external communication unit or through individual departments.

If you have a subject you consider suitable for local, national or international announcement or comment, it is a good idea to discuss it with your science communicator or with the external communication group to obtain their advice on the most effective way to go about it.

Media releases are drafted either by your own science communicator or the external communication group in consultation with you and your colleagues, then cleared by the relevant officers in your department and any research partners. As timing can sometimes be crucial, your help in keeping the clearance procedure swift and accurate is appreciated.

Scientific opinion

There will always be issues on which research staff hold differing scientific opinions. This fosters healthy debate and helps the organisation to develop a balanced position. However, the community sometimes expects a uniform institutional view on a topic and it may occasionally be desirable to try to reconcile widely differing scientific opinions or else form a consensus as to their merits.

Good internal communication, including close collaboration between managers, is essential in helping to resolve differences of view or deciding how best to explain them. Public debate in the media over differing views should be part of a planned strategy to help the community understand that a range of scientific opinion exists and that further research is required to clarify uncertainties or determine where the weight of evidence lies.

Respect for the work of colleagues is very important. It is sensible to make yourself aware of the work of other researchers (and those of partner institutions) who may be involved in the same issue and to check with your research leader or a communicator before committing yourself to an important public statement.

If the issue is sensitive or controversial, please inform your research leader or communicator about your intended public comments. This enables them to provide supporting comment if approached independently by the media, and also to brief you on aspects of the issue and how it affects the organisation of which you may be unaware.

Where you are aware of diverse views on your topic within the organisation, courtesy as well as common sense makes it advisable to let the other researchers know your intentions.

Partnerships and commercial collaboration

Modern science involves extensive partnership and collaboration with many other public and private research institutions, government agencies and private sector companies. While we are firmly in favour of disclosure and knowledge sharing as a general principle, it is important to recognise that there may be legal limitations and contractual obligations as well as the feelings of partners to be considered.

Take care not to disclose unauthorised information about a company, government agency or other organisation working in partnership with our organisation or which has signed a contract with us to do research. Its release might cause embarrassment or financial loss to the other party, may constitute a breach of contract and may harm our standing as a trustworthy provider of research.

Should you be asked to comment publicly on the activities of any commercial or partner organisation, it is advisable to contact that organisation before making comment and seek their reaction to the request. Ascertain whether what you may disclose is subject to a legal confidentiality agreement.

Should the partner not wish you to disclose facts which you feel we have a duty to disclose in the public or national interest, discuss the matter with your research leader and/or senior management. Please don't simply disclose those facts unless you are authorised to do so. The organisation feels strongly its obligation to the public interest and also to its research partners. Balancing these requires tact and careful consideration.

Non-scientific issues

Public statements on non-scientific issues are sometimes made by our scientists as a part of their duties. This responsibility will be given to you by your manager, who is required to brief you on all relevant issues and keep you informed of important developments. You are accountable to your manager for any such statements made, and responsible for informing colleagues who need to know what has been said.

This organisation provides objective scientific information and advice. Though it may contribute to the formulation of policy by government in an advisory capacity, it does not publicly comment on the policies adopted by the government, the opposition or other political parties.

Personal comment

If you wish to publicise your own views on an issue but are not authorised by the organisation, you may do so freely as a private individual. However, you must state plainly that the opinion you give is a personal one and not an official or unofficial view of this organisation. You are asked to help the media to understand the distinction.

Staff whose duties include advising on, or implementing, aspects of government policy should avoid public comment which might conflict with those duties.

If you have any doubts or concerns about expressing a personal opinion, and whether it may be interpreted as an official view, you are encouraged to run them past your communicator or manager and seek their advice. It is also advisable to inform your manager if your views are likely to stimulate public debate or provoke controversy.

Senior managers need to take particular care when making public comment in a private capacity as, despite their insistence that they are speaking privately, they may not be able to escape identification with this organisation.

Please don't use organisational letterhead, envelopes, fax headers or email systems for correspondence in which you express private opinions. The use of any form of the organisation's name, logo or livery will convey an impression your comments are official and authorised.

Public inquiries

Staff are at liberty to make personal submissions to public inquiries with the same qualifications that apply to public comment: do not disclose confidential information without authority, consult your manager and make it clear that your views are privately held. Don't use official stationery or equipment.

Official submissions, which address matters on which the organisation has acknowledged expertise and authority, should be handled by the senior manager responsible for your department, or through the Chief Executive's office.

If you are asked to appear as a witness before a Parliamentary Committee, contact the person responsible for government negotiations for advice that will assist and protect you.

External bodies

Our staff are often asked to serve on external bodies such as committees of inquiry and reviews of other bodies or laws, or as members of community organisations. First, establish whether you are being invited as a representative of the organisation or as an expert individual.

If you represent the organisation, make sure so far as possible that your comments are consistent with official policy, corporate and scientific knowledge on that topic. The task may call for careful differentiation between formal policy, a consensus position among our staff and the need to use your own professional judgement. As you represent the whole organisation, you should consult colleagues who may be able to contribute information and advice that will help you.

If the external body wants you in a personal capacity, make your private status quite clear and insist that nothing the external body says or publishes can be attributed to the organisation. You should also notify your research leader or manager.

Your first point of contact on any aspect of public comment is your Department's Science Communicator. If you would like advice from our External Communication Office, they may be contacted at ...

The document then goes on to list the various laws, statutes and codes of practice to which employees are subject, and to explain briefly their obligations under each. These include:

- any restrictions imposed by the enabling Act or charter of the organisation on public comment
- government or public service rules on public comment, where the organisation is a government statutory authority and its scientists technically public servants
- national security restrictions that may limit public comment on certain issues and impose penalties on those who break them
- privacy laws that may prevent the organisation from disclosing private and personal details of individuals involved in its research activities
- terms and conditions of service imposed by the organisation and agreed to by individuals at the time of their employment
- defamation and libel laws
- commercial and other contractual agreements with research partners
- intellectual property and copyright laws
- scientific ethics codes
- the organisation's code of external business practice.

The most convenient way to disseminate public comment guidelines is via the intranet, if this reaches all staff, or else through an inexpensive booklet that is issued to all staff for easy reference. The guidelines must be regularly reviewed to take account of organisational change as well as changes in the external context, such as new laws and regulations.

Striking a balance between a scientific institution's natural desire for freedom of speech and its obligations to an increasingly complex network of partners, stakeholders, funding agencies, government departments and legal requirements is an extremely delicate business. It is worth bearing in mind that the imposition of authoritarian rules and restraints on scientists' freedom to comment publicly can not only cause resentment but may actually backfire if individuals decide to challenge or ignore them. Appeals to common sense, courtesy and loyalty to one's colleagues, along with a consultative approach, are preferable.

The important thing from the standpoint of open science and human progress is to preserve scientific freedom of speech and to keep public comment as unrestricted as possible. It is also essential for researchers to know they are encouraged to communicate, will be rewarded for doing so and will not be punished or persecuted if they do it in a sensible way. Most publicly funded science is withheld from the public, not because its disclosure is forbidden, but because it is surrounded by a web of conditions, restrictions, obstacles, costs and implied threats – and this often deters scientists from sharing their knowledge.

Finding this web too tangled and complex to negotiate, many researchers decide it is simply not worth the bother of sharing their science outside their peer group. Scientists who lack confidence in their public communication and media skills, or who view communication as a diversion from their research goals, sometimes invoke this web of restrictions as an excuse for avoiding their public duty to communicate. Where scientists fail to communicate externally, they lose the benefits of public or industry feedback on their work. This exposes them to the risk that the science will fail to meet community standards or needs, or that its adoption or commercialisation will be retarded or fail.

It is not in the interests of either science or society for knowledge sharing to be impeded. Clear guidelines are one way that the senior management of a scientific institution can send an unambiguous signal to staff that communication is an approved and valued part of the process of knowledge sharing; indeed it is a duty. However, it is also important for these guidelines to be negotiated and agreed in a transparent and consultative process within the institution, or else staff will feel little ownership of them and have little reason to feel bound by them.

It is also desirable that reward, recognition and promotion in scientific institutions more strongly reflect the duty to share knowledge with the wider community than is usually the case. For scientists to be rewarded or promoted solely on the basis of research, discovery and scientific publication devalues the part of the knowledge system that actually makes their work of worth to society, and is a strategic error.

Forward-looking research institutions are now experimenting with new ways to honour, reward and motivate staff who show a commitment to knowledge sharing. These include:

- annual performance evaluation and key performance indicators that are specifically related to science communication, as distinct from purely research activity
- inclusion of a 'duty to communicate' clause in the employment contract and job description of every researcher, especially senior research managers
- permitting staff to retain income earned from external speaking engagements, media appearances, articles and the like
- requiring a communication plan for every significant research program as a condition of funding support
- including an 'agreement to publicise' clause in commercial research contracts
- providing free training in media skills, public presentation and science writing to staff
- nominating staff for external awards and distinctions that recognise the value of good science communication
- creating internal awards, of equal status with scientific awards, specifically for science communication/knowledge sharing
- recognising the value that professional science communicators can contribute, not only to the organisation but in helping individual scientists to be more effective communicators of their work
- resourcing science communicators on an equivalent scale to researchers, in recognition of their important role in sharing knowledge with society.

Chapter 12

Issues management for science bodies

An organisational 'crisis' is an event that injures the good name, level of trust or scientific credibility of the institution with its stakeholders, customers, partners, peers and the general public, and limits its ability to do its job. It may also have a wider impact: adversely affecting an industry, the community or even the nation.

In most cases, it is an outcome of poor planning and bad management on the part of the research organisation and not, as is so often thought, purely a consequence of attacks by malicious critics or a mischievous media.

An 'issue' is anything that has the potential to affect the performance or reputation of the organisation, either positively or negatively. Not all 'issues' need to be bad news. In some cases, positive issues can be managed in a way that brings greater credit to the organisation and enhances trust in its work.

However, an issue becomes a crisis when, through a failure to anticipate and manage it properly, it goes out of control in a seriously negative fashion, or when an unanticipated internal or external event precipitates it. There are generally three types:

- sudden crises (e.g. natural disasters, fires or major accidents)
- emerging crises (e.g. a build-up of adverse comment in the media, leakage of sensitive information or festering staff disputes)
- chronic crises that run for long periods, sometimes years, with periodic flare-ups and lulls.

Good management of issues and crises involves first the identification of such issues, including the ability to understand how they are likely to be perceived by external audiences, and then the development of a plan to manage the issue. The basic philosophy is that an ounce of prevention is better than a ton of cure.

Most private companies have an issues and crisis management strategy and devote significant staff and financial resources to anticipating and avoiding problems. They recognise that a poorly handled issue can cost millions of dollars and destroy or undermine the good name of the organisation with its clients and society. Most scientific institutions, on the other hand, cling to the wish that 'it'll never happen to us', (in spite of ample evidence that it happens quite often). This is partly due to the scientific conviction that a dollar spent on 'administration' is a dollar wasted and to the lack of a realistic appreciation how damaging to the organisation's scientific credibility a badly handled issue can be – at least in the eyes of the public, politicians, stakeholders and funding sources. Consequently, scientific bodies experience quite a lot of crises, are generally taken by surprise when they break out, and are forced adopt a fire-fighting approach to their control and tend to manage them poorly and reactively.

Issues affecting scientific institutions tend to arise in four main areas:

- scientific and technical
- commercial and legal
- organisational and policy
- people, health, industrial relations, safety and property.

In planning to avoid a crisis, it is sensible to evaluate every unit or department according to this framework in order to be reasonably confident of anticipating what may go wrong.

The best way to do this is for each unit to have its own issues and crisis management (ICM) team that scrutinises the full portfolio of activities regularly and prepares a brief summary of each issue identified. It is crucial to enlist staff support for this process. One of the most effective ways is to explain to all staff what an issue is and why it is important to manage it, then to provide them with the contact details of key members of the ICM team and ask them to report to these individuals any issue they consider to be of concern. This way the staff share responsibility for upholding the organisation's public reputation and feel engaged in the process without being burdened by large administrative responsibilities.

Unit ICM teams then provide regular reports on their most prominent issues to the organisation's overall ICM team. This typically consists of the director, vice-chancellor or chief executive, the chief legal adviser, the human resources manager, the chief communicator and the manager(s) of the area in which a particular issue has arisen or that is affected by it.

Once an issue has been identified and defined, a plan is drawn up to manage it. This plan may involve ongoing active management by many people, or it may simply be a document that sits in the bottom drawer for use in the event that the issue escalates.

The two golden rules of issues management are:

- the public interest is paramount
- there is an absolute commitment to openness and honesty.

These are easy to say but, in science (or anywhere else for that matter), can be unbelievably difficult to follow in practice. The first is often broken because institutions perceive their own interest and welfare as their first priority, not realising how much damage they do to it by being seen to place it above the public interest. Typically, this takes the form of refusal to comment and all the usual behaviour involved in the classic cover-up. Many institutions experience difficulty in differentiating between their own view of what is good and the public good. To follow this rule requires the discipline of managers being prepared to abandon their institutional focus and view the issue from the external perspective of the community and stakeholders. Failure to do this can leave a lasting public impression that the institution is narrow, selfish, biased, arrogant and unworthy of public support or credence. In extreme cases, it may lead to the withdrawal of public sanction for it to perform science and to legal prosecution.

Another reason it is hard for modern scientific bodies to put the public good first is because of binding commitments to commercial and other customers. Nevertheless, a scientific organisation that consistently appears to place the interests of 'big business' (or even 'big government') ahead of the community will suffer loss of trust, credibility and, ultimately, public willingness to adopt its science. This will backfire on its industry customers as, to some degree, happened in the GM food scenario.

An interesting way to test your organisation's capacity to handle issues is for management to pose itself a question like this: 'If, in the course of commercially confidential contract research for a big food company, you discovered food poisoning microbes in their production process, whom would you inform first?'

Those who reply 'the commercial partner' are probably not yet in a frame of mind in which they can successfully manage issues or protect their own reputation. They will be inclined towards cover-ups. The public interest comes first.

The second golden rule of absolute commitment to openness and honesty is equally challenging, especially for those organisations with strong bureaucratic or academic traditions of secrecy. However, few things are more harmful to a reputation than an appearance of contempt for the public's 'right to know'.

As in so many human relationships, honesty is the best policy and truth the safest course. If the organisation has erred in some way, a confession and a plea for public forgiveness are more confidence-inspiring than a botched attempt to 'spin' one's way out of it. The literature on corporate crisis management is filled with case studies that reinforce this principle; indeed, there are many cases where absolute honesty and candour led to the organisation's standing being enhanced, rather than diminished, by the issue.

Of course, organisations do stage cover-ups and get away with them – but any smug self-congratulation at having successfully 'handled' the issue and put the media off the scent is likely to be short-lived if it becomes a habit. Not least of the reasons for this is that all organisations contain honest and decent staff who are likely to be morally offended by a cover-up and who may therefore be strongly inclined to 'blow the whistle' or leak to the media. Scientists, more than most professionals, feel compelled to speak out if their values are affronted.

One of the best ways to view an issue is to assume from the start that it will eventually become public knowledge, and to plan the response accordingly.

Here are some basic steps for establishing an issue management plan.

ICM team

Decide who is to be on the ICM team. The members must be prepared to drop all current commitments and responsibilities to focus totally on the issue. List all their 24-hour contact details.

Short description

Prepare a concise written description of the issue, explaining clearly what it is about and its likely impact on the organisation, its partners, the government and community.

Our policy

Draft a succinct statement of the institution's position on this issue. If none exists, then develop a clear policy.

Spokesperson: 24-hour contact

Nominate who is to be the lead spokesperson on the issue. This individual will be very senior (often the CEO), understand the issue well, be articulate, and have a reassuring demeanour and the ability to stay cool under fire. Preferably they will have experience and/or training in how to respond to media questioning. This individual will be the public face of the institution while the issue is running and must be contactable around the clock.

The public position

Summarise the issue and what the institution is doing about it, in plain language and dot point form. Use an experienced communicator to draft this document, which is for use by the spokesperson, other staff who may be asked to comment, and to brief stakeholders. It will also be used to respond to media and public inquiry. It will need to represent the issue from the public interest, as well as the scientific, perspective.

Legal obligations

The ICM team must methodically consider each aspect of the issue from the standpoint of the institution's legal and contractual obligations. It is common for the legal adviser to be a key member of the ICM team. Beware of allowing legal considerations to overwhelm the need to be open and honest with the public, however.

Stakeholders

The ICM team must prepare a plan for informing key stakeholders about the issues, and whether it affects them directly or indirectly. These approaches are best made in person to gauge the response of the partners and whether they are likely to help or hinder effective management of the issue.

Other managers

The heart of good issues management is good internal communication. The ICM team always includes several senior managers and, often, the head of the organisation. However, it is essential that other managers be kept in the loop with regular updates and consultation, and can feed in their advice readily. Any of them could face media or stakeholder demands for comment. They must also be in a position to brief and reassure their own staff.

Government

For public research institutions, universities and companies that produce consumer goods, it is essential to keep government fully briefed on the situation, in particular the office of the minister responsible. Politicians hate surprises – and crises can be used by political opponents to mount an ambush.

Tell the staff

A plan for informing the staff is essential. This must swing into effect as soon as the issue becomes, or threatens to become, public. Informed staff are likely to be supportive and helpful; staff kept in the dark by a thoughtless, secretive management are likely to feel resentful and angry, and may be tempted to add their voices to external criticism. The media often gets its leads from concerned staff, or from chatting with contacts who work for an organisation. A journalist who has run into a brick wall of 'no comment' may turn to the switchboard operator, the canteen staff or a former schoolmate for leads to follow up. It is preferable for management to make the public statements, rather than the janitor.

Public response

The ICM team prepares a step-by-step plan for responding to public concern and inquiries, and ensures that staff clearly know what their role is and that adequate resources are available. The response must take first account of the public interest (e.g. safety, health and jobs) as opposed to the institutional interest. It must be easily accessible through hotlines, email, RSS, blogs and web services, public briefings, media announcements, and so on.

Media response

Identify key media and journalists to be contacted and briefed on the issue. An informed media that is taken into the organisation's confidence will often prove more an ally than a foe. Spin, silence and an uncooperative stance, on the other hand, are read by the media as signs of guilt and an invitation to dig deeper. In dealing with journalists it is vital that they do not feel they are being manipulated or used to deliver organisational propaganda: they will certainly be thinking first of the public interest. How open and honest the organisation appears will often decide how they report on its handling of the issue.

In a major or long-running crisis, it is sound practice to commission opinion research, media and web analysis to track changes in public sentiment, and identify those wounds from which you are still bleeding. This will also enable you to monitor how particular media and journalists treat you over time.

HANDLING ADVERSE REACTION

If there is a strong or violent public reaction to a scientific organisation, even if it is confined to small groups within the community, it is wise to consider the following possibilities.

Attacks on staff

Threats, verbal abuse and even physical attacks may be aimed at the staff of a scientific institution by hostile individuals and groups. Receptionists are often targets for public ire, as are senior managers and those named as being involved in controversial work. Threats may be delivered by email, mail, phone, the internet or in person.

A system for logging all such threats and arranging physical or police security for affected staff must be put in place that can operate at very short notice. Phone, electronic and written threats should be preserved, if possible, to assist police inquiries. Staff who are targets of these attacks, or even witnesses to them, may be distressed and require counselling.

Attacks on property

Hostile groups may also attack, damage and deface scientific property, particularly if it is not well protected. Sites, vehicles and other property bearing corporate signage are especially vulnerable and ought not to be over-exposed. Property security and surveillance should be reviewed and increased if warranted by the threat or scale of external actions.

Protests

Physical protests and disruption of scientific events on- and off-site must also be considered. These can range from harmless but embarrassing incidents (such as nude demonstrations or pie throwing) to highly confrontational and violent incidents. Security at public, official and media events should be reviewed. Indeed, thought should be given to postponing or cancelling events that offer particular opportunity to protestors.

Local residents may become concerned and angry if a research site in their neighbourhood is targeted by protesters. This may add to the overall volume of external criticism. Steps should be taken to brief local residents and seek their feedback and support. In the long term, their tolerance will be critical to the research organisation's presence in the locality.

Electronic attacks

Contemporary protest may take the form of attacks on email and websites, unauthorised break-ins, computer viruses, distribution of offensive material to staff and organised crashing of IT systems. Close attention must be paid to electronic security, as well as considering what information may be on public display that might inflame negative sentiment.

Political flak

Explanations of the research and its implications may be called for by government ministers and officials, and should be ready in advance. Ministers like to be briefed of impending issues before they become public.

Oppositions may exploit scientific issues to attack the government. Where permitted, it is sensible to brief both sides (government first) to reduce the chances of becoming a 'political football'. On some occasions, local government may also need to be briefed.

Public funding consequences

There may be a loss of public confidence, and hence public funding, as a result of prolonged adverse publicity. A plan to brief and shore up trust among funding sources is highly advisable.

Stakeholder consequences

Consideration should be given to all external stakeholders, including those not directly connected with the research in question, who may become uneasy or hostile as a result of prolonged bad publicity. Steps may be needed to reassure and update them. These include commercial partners and other research institutions.

Staff consequences

In managing a crisis, the trust, support and cooperation of staff are of paramount importance. So is keeping them fully informed. Nothing is more damaging than the deliberate leaking of information or rumour-mongering by disillusioned staff who feel ignored, mistrusted or kept in the dark. Junior staff are as important as senior officers – the switchboard operator may receive far more media interrogation and angry public reaction than the director.

Staff can also assist by acting as a sounding board and a conduit for public opinion. Counselling may be needed for the distressed.

Peer consequences

Harsh judgements and criticism of the institution responsible for the research are likely from scientists and academics in the same, as well as unrelated, fields, which leads to a generally negative professional view of the organisation. The scientific standing of the organisation will need to be considered, and steps taken through professional bodies and scientific publications to restore trust, if needed.

Alienation of supporters

Community groups, industry and non-government organisations that are normally supportive of scientific research may become hostile and angry if they form a view that the work has developed outcomes adverse to their beliefs or interests. They will need openness and reassurance.

Ethical objections

In a serious controversy, public objections to the research may be expected on religious, moral and ethical grounds, which need to be answered on those grounds. In particular, it is important to explain what sort of ethical oversight there was and to demonstrate that community values are heeded.

Email and web discussion groups

A lot of nonsense and scaremongering is peddled on international email and web 'blogs', and may become a source of self-sustaining attack. It is highly advisable to monitor this discussion and, even at the expense of personal criticism, to seek from time to time to inject the facts, simply and without emotion. Monitoring popular blogs and opinion sites is also a good way to anticipate fresh issues before they become widely established, and to prepare for them.

Future conduct of research

A serious public controversy over a piece of science has implications for the future conduct of the research, and of research like it. Issues to be considered include the extent of public oversight and external and government review of the work and its progress. Where public trust is damaged, there may be strong political pressure for termination of the work in question.

Damned lies

One of the hardest issues to deal with is the promulgation by critics/opponents of outright falsehoods or cleverly exaggerated claims. If these occur, the response needs to be:

- calm
- totally truthful and based on fact

- always referring to the best interests of the public and community (not those of the institution or the researchers)
- presented by an experienced and media-savvy senior officer available for comment 24 hours a day
- endlessly repetitive
- based on an agreed set of responses that all those involved have, know, support and understand
- short, clear and crisp to meet media requirements.

In managing issues, there is no substitute for preparedness. The entire research portfolio should be regularly reviewed for any work with the potential to generate hostile public reaction, even from small segments of the community or lobby groups. In every case identified, a plan must be devised for handling the issue when (not if) it breaks.

In science, a problem is that researchers so love their work that they find it hard to imagine that anyone might see it as dangerous or unethical, or else they may feel it is so important that mere public criticism ought not to be allowed to stand in its way. Consequently, they make poor judges of its potential to stir up hostility in the wider community. Even when they are aware that their work is controversial, there is still a great tendency to hope that 'it will never happen' or to trust that the organisation's good name will see things through. As scientists are often ill-acquainted with the media and its methods, they may also under-rate the seriousness of an issue and how quickly it can spiral out of control.

These are the main reasons why research and academic organisations are poor at issues management. It is therefore very important that the potential of a research project to blow up into an issue or crisis is evaluated by someone who can bring an external perspective – a science communicator, journalist, public relations expert, consumer lobbyist or professional issues manager – even an experienced business person or politician.

Eleventh-hour attempts to salve a 'crisis' can do no more than slightly mitigate the resulting damage to the organisation's reputation. These attempts are no substitute for early identification and planning.

The most important cultural change needed in a scientific institution wanting to improve its ability to manage issues is to have both management and staff understand that crises are generally an outcome of poor internal practices – not a consequence of attacks by their critics or the media. Shooting the messenger seldom obliterates the message.

After a crisis

After a crisis it is important for an organisation to understand clearly how much damage it has sustained and how this may have affected its standing with key stakeholders, clients, the public and its own staff. This can be done through quantitative and qualitative opinion research and by media analysis, as previously described.

The next step is to consider how best to repair any damage that may have been sustained to the organisation.

If damage exists among stakeholders and partners, then a methodical process of rebuilding mutual trust and confidence must be set in train. This will invariably involve re-establishing or strengthening personal contacts between the bodies. A series of meetings between managers on both sides is important to understand the nature of the stakeholder's reaction to the issue. Is it a loss of faith in scientific or managerial professionalism, in integrity, or a resentment of having been publicly linked with an unsavoury episode? Only when this understanding is clear can effective efforts be made to re-establish the relationship.

Scientific organisations commonly use the 'deficit model' of public reputation, although not always consciously. In this model, the reputation of the organisation is treated as if it were a bank account with a whole series of positive deposits (reported beneficial achievements) over time adding up to a healthy balance. Along comes a crisis and there is a sudden draw-down in the organisation's credit. If the crisis is serious enough, the account goes into the red, meaning that the overall external perception of the institution is a negative one. Even very ancient and august establishments, such as the great old universities, are no more immune from a negative balance of public perception than an ancient banking house is immune from bankruptcy.

This model must not replace proactive issues management as a way of dealing with negative sentiment, because good planning can dramatically reduce the extent of the withdrawal of one's credit balance.

However, where the institution is seen to have suffered a loss of credit, great or small, then the intelligent course is to redouble its efforts to get some positive information back into the public arena. This may involve listening carefully to the tone of public criticism, and providing examples of the organisation's recent achievements that will offset it.

The 10-to-1 rule says that if you have had 10 bad stories published in the media about you, you will need to aim for 100 positive stories to offset the impact. In practice, it may take several years of consistent hard work and positive exposure to overcome the negative impact of a particularly serious issue on your credit balance. The full cooperation of your scientists and communicators will be needed to increase organisational output of information beneficial to the public.

This may sound a little self-serving, but one of the positive things about having a crisis is that it prompts a certain amount of internal questioning and this may lead to a recognition of the need to be more open, communicative and heedful of public opinion and to factor public benefit into research planning more consistently. External communication then becomes more closely attuned to the public's needs and interests, rather than the organisation's, and so becomes a more effective form of knowledge sharing.

As important as addressing external trust is the need to repair any internal loss of confidence that may have occurred. After a crisis, extra effort is required to convince sceptical staff that management has listened and learned from the event and is prepared to change its behaviour where needed.

IMAGE AND BRAND MANAGEMENT

There is much confusion among scientific and academic institutions about the need for corporate image or brand management. Many bodies fall prey to commercial spin-doctors and image-merchants who charge hundreds of thousand dollars for a report telling the institution what it probably knew, collectively, in the first place and that will leave it little the wiser as to how to go about improving the situation (unless, of course, it hires the image crafters at a further massive cost).

The confusion arises from the fact that universities and research agencies are engaging in more and more contract work for the private sector and are selling more of their 'products'. They are learning the language, mode of thought and ways of commerce as they get closer to their customers. This is laudable and necessary. Where error creeps in is in the assumption that what is right for commerce in general is also right for science or academia.

It is, of course, essential for a scientific or academic body to have a high public reputation, a clear identity and a prominent and trusted brand. These are at the heart of its ability to attract good staff, plentiful funding and to achieve the ultimate adoption or commercialisation of its research. If it equates itself with a commercial organisation, the best an institution can hope for is a good commercial reputation, which enjoys nothing like the public confidence reposed in public institutions.

Truth beats fiction

The core issue is whether the image is based upon genuine merit and unique attributes or whether it is a synthetic one derived from publicity, spin-doctoring, advertising, self-promotion and other highly commercial techniques. The risk in the latter course is that the public are not fools, and soon spot a 'show pony' from a genuine stayer.

In commerce, whether you make breakfast cereal, soap powder or toothpaste or deliver banking services, you are one company among many offering similar products. It is vital to try to differentiate your product from your competitors', even if it isn't really all that different. That's where good marketing, PR and corporate image-building come in.

The situation is different for science and academia. Although scientific and academic institutions may perceive themselves as being in commercial competition with one another for research work, students or funding, in reality their products are usually quite distinctive. Scientific research, in particular, is unique because that is the special feature of scientific discovery: peer review and reporting. Research that isn't unique is plagiarism.

The principle is that the corporate image or brand of a scientific institution rests not on artificial promotion but on the communication of its genuine achievements and real contribution to society over time.

The best way for the organisation to build and sustain its image is to communicate the outcomes of its research and their value to society or industry as effectively, comprehensively and openly as possible. There is no need to embellish: scientific excellence speaks for itself, although the details may require clear explanation to some audiences.

Some institutions forget this and embrace the techniques that work for breakfast cereals, soap powder, toothpaste or banking services. Apart from being very costly, the use of these techniques does not inspire the public with confidence in a research institution – they are the tricks used by commerce to purchase market advantage, and their use in science simply makes the institution look shabby or desperate.

Listening skills

The second key ingredient in the successful image management of a scientific organisation is to have highly developed listening skills that enable it to hear and understand what is going on in the world outside. Techniques such as 'reading the public mind' are invaluable in this.

In the traditional model – far too commonly, alas – the organisation goes about its business with absolute concentration and focus, ignoring the outside world and then being surprised when it finds itself out of step or facing a crisis. Like a piece of emergency equipment in a glass case, the communicator is assigned a subsidiary role – 'do what the scientists tell you' – rather than being a key part of the management team advising and planning the external profile. Classically, if there's a crisis, the organisation smashes the glass case, tells the communicator to get out and 'fix it', and then blames them when the attempt fails.

The main failure here is one of research. Not enough is known about what is happening in the outside world, and the institution has become out of touch and fallen behind changes in politics or community expectations, standards and demands. Essential to having the world think well of you is to understand what the world sees when it looks at you. Almost certainly, it will be different to what you imagine.

George Littlewood, an internationally experienced public affairs manager for the mining industry, says that one of the reasons the miners came in for so much criticism in the late 20th century was that they were slow to appreciate how people's understanding of, and care for, the environment had grown. The miners had assumed the public mainly wanted them to create wealth and jobs, so they didn't need to bother too much about issues such as relations with the community, native peoples and the environment. It wasn't until they did detailed qualitative opinion research that the miners discovered, with shock, that the public wanted them to do better in *all* areas, *at the same time*.[1]

'Community and government expectations will change over time, and there is a danger for your organisation in your performance shifting too far from their expectations,' Littlewood says. This creates what he terms a 'legitimacy gap', which falls between the organisation and the public sanction it has to do scientific research. This will require the organisation to improve the way it communicates with and engages various audiences and to change its behaviour so it is in tune with changing community expectations and requirements, and understands them better.

Obscure and crazy names

When it comes to choosing a name, many scientific organisations shoot themselves in the foot. They pick a name that describes themselves to colleagues, but which is mysterious, opaque or plain misleading to the outside world. And then they complain the media ignores them.

The identity of a scientific institution is comprised of four things: its purpose, achievements, name and imagery.

A poor choice of institutional name not only hampers public recognition and reputation – but also the things that derive from this, such as funding, the uptake of science, public and political influence. It also conveys an impression the organisation is out of touch.

Many scientific bodies are furious when the media leaves their name out of a report – without realising that they set themselves up for it when they chose a long, obscure name. The media isn't interested in wasting valuable time or space on long, clunky names. It is far more concerned with the substance of the story and what it means to its readers or viewers.

If yours is a new scientific organisation and you have the privilege of devising its identity, name and symbols, then a sensible course is to consult with the likely customers and stakeholders as well as the staff and management. If, on the other hand, the organisation carries a long and burdensome title from yesteryear that for various reasons – both reputational and sentimental – it is unwilling to relinquish, it is still possible to coin a working title, a public brand or slogan that better explains what the organisation actually does.

A good example of this is the World Bank. Its actual name is the International Bank for Reconstruction and Development (IBRD), but everyone calls it the World Bank because it is short, neat and conveys the idea that it is global and has something to do with money. Another example is a marine research institute, the International Centre for Living Aquatic Resources Management (ICLARM), which wanted to raise its public profile without abandoning a respected professional name – so it selected a brand, *WorldFish*, to use in conjunction with the formal name. Its global media profile jumped immediately.

Just as parents can curse a child for life with a poor choice of name, the same is true of scientific organisations. It is important for a name to reflect the values of those within an organisation – but it should also be a name that has meaning and value to those outside, especially if they are paying for it. The best plan is to test internally proposed names out on various audiences outside the institution using focus group or similar techniques, to see how they react not only to the full title but also to its component words, as well as what images the name conjures up. This can sometimes be quite a shock!

This also applies to logos. What may be aesthetically appealing, or a clever embodiment of key institutional elements, can also create public mystification and confusion – especially if it has unfortunate connotations outside science. A logo that means something to the outside world is more likely to work than one that is obscure or just arty. Good design always embodies function as well as form.

'Branding' a science institution is therefore not quite the same as branding a company or commercial product. It needs to be carried out with great care and thought for the public interest, which it should always convey.

AFTERWORD

In the era of 'peak everything' – when human needs are coming into constant collision with the Earth's available resources – the need for open science is paramount.

Since its earliest beginning, science has advocated openness, but a 20th century of mistrust, suspicion and outright hostility among nations quenched much of its commitment to this essential value. Science has been conscripted to serve the ends of military and economic domination, and its dealings with society in general restricted accordingly. In many cases it has been gagged, which, as Sir Henry Dale and Sir David Rivett pointed out so long ago, is bad for science and bad for society.

Science itself flourishes best when there is no hindrance to the free exchange of knowledge. Societies flourish the most when science and its resulting technologies can be transmitted, examined and deployed most fruitfully and rapidly. This book *Open Science* has explored many tried and tested ways these desirable outcomes can be achieved in the hope they will be more widely employed within science itself, industry, government and human society at large.

As the collision between human demands and the Earth's resources becomes more acute, and the risk of catastrophe magnifies, the importance of an open science in which new and better knowledge flows most swiftly to those who most need it will be overwhelming.

We end this book with an appeal to those who read it to do whatever is in their power to help rebuild and fulfil the ideal of a science open to all humanity, using all the wonderful communication means now available.

APPENDIX: DECLARATION ON SCIENCE AND THE USE OF SCIENTIFIC KNOWLEDGE

Text adopted by the World Conference on Science, 1 July 1999.

PREAMBLE

1. We all live on the same planet and are part of the biosphere. We have come to recognise that we are in a situation of increasing interdependence, and that our future is intrinsically linked to the preservation of the global life-support systems and to the survival of all forms of life. The nations and the scientists of the world are called upon to acknowledge the urgency of using knowledge from all fields of science in a responsible manner to address human needs and aspirations without misusing this knowledge. We seek active collaboration across all the fields of scientific endeavour, that is the natural sciences such as the physical, earth and biological sciences, the biomedical and engineering sciences, and the social and human sciences. While the Framework for Action emphasises the promise and the dynamism of the natural sciences but also their potential adverse effects, and the need to understand their impact on and relations with society, the commitment to science, as well as the challenges and the responsibilities set out in this Declaration, pertain to all fields of the sciences. All cultures can contribute scientific knowledge of universal value. The sciences should be at the service of humanity as a whole, and should contribute to providing everyone with a deeper understanding of nature and society, a better quality of life and a sustainable and healthy environment for present and future generations.
2. Scientific knowledge has led to remarkable innovations that have been of great benefit to humankind. Life expectancy has increased strikingly, and cures have been discovered for many diseases. Agricultural output has risen significantly in many parts of the world to meet growing population needs. Technological developments and the use of new energy sources have created the opportunity to free humankind from arduous labour. They have also enabled the generation of an

expanding and complex range of industrial products and processes. Technologies based on new methods of communication, information handling and computation have brought unprecedented opportunities and challenges for the scientific endeavour as well as for society at large. Steadily improving scientific knowledge on the origin, functions and evolution of the universe and of life provides humankind with conceptual and practical approaches that profoundly influence its conduct and prospects.

3. In addition to their demonstrable benefits the applications of scientific advances and the development and expansion of human activity have also led to environmental degradation and technological disasters, and have contributed to social imbalance or exclusion. As one example, scientific progress has made it possible to manufacture sophisticated weapons, including conventional weapons and weapons of mass destruction. There is now an opportunity to call for a reduction in the resources allocated to the development and manufacture of new weapons and to encourage the conversion, at least partially, of military production and research facilities to civilian use. The United Nations General Assembly has proclaimed the year 2000 as International Year for the Culture of Peace and the year 2001 as United Nations Year of Dialogue among Civilisations as steps towards a lasting peace; the scientific community, together with other sectors of society, can and should play an essential role in this process.
4. Today, whilst unprecedented advances in the sciences are foreseen, there is a need for a vigorous and informed democratic debate on the production and use of scientific knowledge. The scientific community and decision-makers should seek the strengthening of public trust and support for science through such a debate. Greater interdisciplinary efforts, involving both natural and social sciences, are a prerequisite for dealing with ethical, social, cultural, environmental, gender, economic and health issues. Enhancing the role of science for a more equitable, prosperous and sustainable world requires the long-term commitment of all stakeholders, public and private, through greater investment, the appropriate review of investment priorities, and the sharing of scientific knowledge.
5. Most of the benefits of science are unevenly distributed, as a result of structural asymmetries among countries, regions and social groups, and between the sexes. As scientific knowledge has become a crucial factor in the production of wealth, so its distribution has become more inequitable. What distinguishes the poor (be it people or countries)

from the rich is not only that they have fewer assets, but also that they are largely excluded from the creation and the benefits of scientific knowledge.

6. We, participants in the World Conference on Science for the Twenty-first Century: A New Commitment, assembled in Budapest, Hungary, from 26 June to 1 July 1999 under the aegis of the United Nations Educational, Scientific and Cultural Organization (UNESCO) and the International Council for Science (ICSU):
   Considering:
7. where the natural sciences stand today and where they are heading, what their social impact has been and what society expects from them,
8. that in the twenty-first century science must become a shared asset benefiting all peoples on a basis of solidarity, that science is a powerful resource for understanding natural and social phenomena, and that its role promises to be even greater in the future as the growing complexity of the relationship between society and the environment is better understood,
9. the ever-increasing need for scientific knowledge in public and private decision making, including notably the influential role to be played by science in the formulation of policy and regulatory decisions,
10. that access to scientific knowledge for peaceful purposes from a very early age is part of the right to education belonging to all men and women, and that science education is essential for human development, for creating endogenous scientific capacity and for having active and informed citizens,
11. that scientific research and its applications may yield significant returns towards economic growth and sustainable human development, including poverty alleviation, and that the future of humankind will become more dependent on the equitable production, distribution and use of knowledge than ever before,
12. that scientific research is a major driving force in the field of health and social care and that greater use of scientific knowledge would considerably improve human health,
13. the current process of globalisation and the strategic role of scientific and technological knowledge within it,
14. the urgent need to reduce the gap between the developing and developed countries by improving scientific capacity and infrastructure in developing countries,
15. that the information and communication revolution offers new and more effective means of exchanging scientific knowledge and advancing education and research,

16. the importance for scientific research and education of full and open access to information and data belonging to the public domain,
17. the role played by the social sciences in the analysis of social transformations related to scientific and technological developments and the search for solutions to the problems generated in the process,
18. the recommendations of major conferences convened by the organisations of the United Nations system and others, and of the meetings associated with the World Conference on Science,
19. that scientific research and the use of scientific knowledge should respect human rights and the dignity of human beings, in accordance with the Universal Declaration of Human Rights and in the light of the Universal Declaration on the Human Genome and Human Rights,
20. that some applications of science can be detrimental to individuals and society, the environment and human health, possibly even threatening the continuing existence of the human species, and that the contribution of science is indispensable to the cause of peace and development, and to global safety and security,
21. that scientists with other major actors have a special responsibility for seeking to avert applications of science which are ethically wrong or have an adverse impact,
22. the need to practise and apply the sciences in line with appropriate ethical requirements developed on the basis of an enhanced public debate,
23. that the pursuit of science and the use of scientific knowledge should respect and maintain life in all its diversity, as well as the life-support systems of our planet,
24. that there is a historical imbalance in the participation of men and women in all science-related activities,
25. that there are barriers which have precluded the full participation of other groups, of both sexes, including disabled people, indigenous peoples and ethnic minorities, hereafter referred to as disadvantaged groups,
26. that traditional and local knowledge systems, as dynamic expressions of perceiving and understanding the world, can make, and historically have made, a valuable contribution to science and technology, and that there is a need to preserve, protect, research and promote this cultural heritage and empirical knowledge,
27. that a new relationship between science and society is necessary to cope with such pressing global problems as poverty, environmental degradation, inadequate public health, and food and water security, in particular those associated with population growth,

28. the need for a strong commitment to science on the part of governments, civil society and the productive sector, as well as an equally strong commitment of scientists to the well-being of society,

Proclaim the following:

1 Science for knowledge; knowledge for progress

29. The inherent function of the scientific endeavour is to carry out a comprehensive and thorough inquiry into nature and society, leading to new knowledge. This new knowledge provides educational, cultural and intellectual enrichment and leads to technological advances and economic benefits. Promoting fundamental and problem-oriented research is essential for achieving endogenous development and progress.
30. Governments, through national science policies and in acting as catalysts to facilitate interaction and communication between stakeholders, should give recognition to the key role of scientific research in the acquisition of knowledge, in the training of scientists and in the education of the public. Scientific research funded by the private sector has become a crucial factor for socio-economic development, but this cannot exclude the need for publicly funded research. Both sectors should work in close collaboration and in a complementary manner in the financing of scientific research for long-term goals.

2 Science for peace

31. The essence of scientific thinking is the ability to examine problems from different perspectives and seek explanations of natural and social phenomena, constantly submitted to critical analysis. Science thus relies on critical and free thinking, which is essential in a democratic world. The scientific community, sharing a long-standing tradition that transcends nations, religions and ethnicity, should promote, as stated in the Constitution of UNESCO, the 'intellectual and moral solidarity of mankind', which is the basis of a culture of peace. Worldwide cooperation among scientists makes a valuable and constructive contribution to global security and to the development of peaceful interactions between different nations, societies and cultures, and could give encouragement to further steps in disarmament, including nuclear disarmament.
32. Governments and society at large should be aware of the need to use natural and social sciences and technology as tools to address the root causes and impacts of conflict. Investment in scientific research which addresses them should be increased.

3 Science for development

33. Today, more than ever, science and its applications are indispensable for development. All levels of government and the private sector should provide enhanced support for building up an adequate and evenly distributed scientific and technological capacity through appropriate education and research programmes as an indispensable foundation for economic, social, cultural and environmentally sound development. This is particularly urgent for developing countries. Technological development requires a solid scientific basis and needs to be resolutely directed towards safe and clean production processes, greater efficiency in resource use and more environmentally friendly products. Science and technology should also be resolutely directed towards prospects for better employment, improving competitiveness and social justice. Investment in science and technology aimed both at these objectives and at a better understanding and safeguarding of the planet's natural resource base, biodiversity and life-support systems must be increased. The objective should be a move towards sustainable development strategies through the integration of economic, social, cultural and environmental dimensions.
34. Science education, in the broad sense, without discrimination and encompassing all levels and modalities, is a fundamental prerequisite for democracy and for ensuring sustainable development. In recent years, worldwide measures have been undertaken to promote basic education for all. It is essential that the fundamental role played by women in the application of scientific development to food production and health care be fully recognised, and efforts made to strengthen their understanding of scientific advances in these areas. It is on this platform that science education, communication and popularisation need to be built. Special attention still needs to be given to marginalised groups. It is more than ever necessary to develop and expand science literacy in all cultures and all sectors of society as well as reasoning ability and skills and an appreciation of ethical values, so as to improve public participation in decision making related to the application of new knowledge. Progress in science makes the role of universities particularly important in the promotion and modernisation of science teaching and its coordination at all levels of education. In all countries, and in particular the developing countries, there is a need to strengthen scientific research in higher education, including postgraduate programmes, taking into account national priorities.

35. The building of scientific capacity should be supported by regional and international cooperation, to ensure both equitable development and the spread and utilisation of human creativity without discrimination of any kind against countries, groups or individuals. Cooperation between developed and developing countries should be carried out in conformity with the principles of full and open access to information, equity and mutual benefit. In all efforts of cooperation, diversity of traditions and cultures should be given due consideration. The developed world has a responsibility to enhance partnership activities in science with developing countries and countries in transition. Helping to create a critical mass of national research in the sciences through regional and international cooperation is especially important for small States and least developed countries. Scientific structures, such as universities, are essential for personnel to be trained in their own country with a view to a subsequent career in that country. Through these and other efforts conditions conducive to reducing or reversing the brain drain should be created. However, no measures adopted should restrict the free circulation of scientists.
36. Progress in science requires various types of cooperation at and between the intergovernmental, governmental and non-governmental levels, such as: multilateral projects; research networks, including South–South networking; partnerships involving scientific communities of developed and developing countries to meet the needs of all countries and facilitate their progress; fellowships and grants and promotion of joint research; programmes to facilitate the exchange of knowledge; the development of internationally recognised scientific research centres, particularly in developing countries; international agreements for the joint promotion, evaluation and funding of mega-projects and broad access to them; international panels for the scientific assessment of complex issues; and international arrangements for the promotion of postgraduate training. New initiatives are required for interdisciplinary collaboration. The international character of fundamental research should be strengthened by significantly increasing support for long-term research projects and for international collaborative projects, especially those of global interest. In this respect particular attention should be given to the need for continuity of support for research. Access to these facilities for scientists from developing countries should be actively supported and open to all on the basis of scientific merit. The use of information and communication technology, particularly through networking, should be expanded as a means of promoting the

free flow of knowledge. At the same time, care must be taken to ensure that the use of these technologies does not lead to a denial or restriction of the richness of the various cultures and means of expression.

37. For all countries to respond to the objectives set out in this Declaration, in parallel with international approaches, in the first place national strategies and institutional arrangements and financing systems need to be set up or revised to enhance the role of sciences in sustainable development within the new context. In particular, they should include: a long-term national policy on science to be developed together with the major public and private actors; support to science education and scientific research; the development of cooperation between R&D institutions, universities and industry as part of national innovation systems; the creation and maintenance of national institutions for risk assessment and management, vulnerability reduction, safety and health; and incentives for investment, research and innovation. Parliaments and governments should be invited to provide a legal, institutional and economic basis for enhancing scientific and technological capacity in the public and private sectors and facilitate their interaction. Science decision making and priority setting should be made an integral part of overall development planning and the formulation of sustainable development strategies. In this context, the recent initiative by the major G-8 creditor countries to embark on the process of reducing the debt of certain developing countries will be conducive to a joint effort by the developing and developed countries towards establishing appropriate mechanisms for the funding of science in order to strengthen national and regional scientific and technological research systems.
38. Intellectual property rights need to be appropriately protected on a global basis, and access to data and information is essential for undertaking scientific work and for translating the results of scientific research into tangible benefits for society. Measures should be taken to enhance those relationships between the protection of intellectual property rights and the dissemination of scientific knowledge that are mutually supportive. There is a need to consider the scope, extent and application of intellectual property rights in relation to the equitable production, distribution and use of knowledge. There is also a need to further develop appropriate national legal frameworks to accommodate the specific requirements of developing countries and traditional knowledge and its sources and products, to ensure their recognition and adequate protection on the basis of the informed consent of the customary or traditional owners of this knowledge.

4 Science in society and science for society

39. The practice of scientific research and the use of knowledge from that research should always aim at the welfare of humankind, including the reduction of poverty, be respectful of the dignity and rights of human beings, and of the global environment, and take fully into account our responsibility towards present and future generations. There should be a new commitment to these important principles by all parties concerned.
40. A free flow of information on all possible uses and consequences of new discoveries and newly developed technologies should be secured, so that ethical issues can be debated in an appropriate way. Each country should establish suitable measures to address the ethics of the practice of science and of the use of scientific knowledge and its applications. These should include due process procedures for dealing with dissent and dissenters in a fair and responsive manner. The World Commission on the Ethics of Scientific Knowledge and Technology of UNESCO could provide a means of interaction in this respect.
41. All scientists should commit themselves to high ethical standards, and a code of ethics based on relevant norms enshrined in international human rights instruments should be established for scientific professions. The social responsibility of scientists requires that they maintain high standards of scientific integrity and quality control, share their knowledge, communicate with the public and educate the younger generation. Political authorities should respect such action by scientists. Science curricula should include science ethics, as well as training in the history and philosophy of science and its cultural impact.
42. Equal access to science is not only a social and ethical requirement for human development, but also essential for realising the full potential of scientific communities worldwide and for orienting scientific progress towards meeting the needs of humankind. The difficulties encountered by women, constituting over half of the world's population, in entering, pursuing and advancing in a career in the sciences and in participating in decision making in science and technology should be addressed urgently. There is an equally urgent need to address the difficulties faced by disadvantaged groups which preclude their full and effective participation.
43. Governments and scientists of the world should address the complex problems of poor health and increasing inequalities in health between different countries and between different communities within the same country with the objective of achieving an enhanced, equitable

standard of health and improved provision of quality health care for all. This should be undertaken through education, by using scientific and technological advances, by developing robust long-term partnerships between all stakeholders and by harnessing programmes to the task.

44. We, participants in the World Conference on Science for the Twenty-first Century: A New Commitment, commit ourselves to making every effort to promote dialogue between the scientific community and society, to remove all discrimination with respect to education for and the benefits of science, to act ethically and cooperatively within our own spheres of responsibility, to strengthen scientific culture and its peaceful application throughout the world, and to promote the use of scientific knowledge for the well-being of populations and for sustainable peace and development, taking into account the social and ethical principles illustrated above.
45. We consider that the Conference document Science Agenda – Framework for Action gives practical expression to a new commitment to science, and can serve as a strategic guide for partnership within the United Nations system and between all stakeholders in the scientific endeavour in the years to come.
46. We therefore adopt this Declaration on Science and the Use of Scientific Knowledge and agree upon the Science Agenda – Framework for Action as a means of achieving the goals set forth in the Declaration, and call upon UNESCO and ICSU to submit both documents to the General Conference of UNESCO and to the General Assembly of ICSU. The United Nations General Assembly will also be seized of these documents. The purpose is to enable both UNESCO and ICSU to identify and implement follow-up action in their respective programmes, and to mobilise the support of all partners, particularly those in the United Nations system, in order to reinforce international coordination and cooperation in science.

ENDNOTES

Chapter 1

1 Meho LI (2007) The rise and rise of citation analysis. *Physics World* **29(1)**, 32–36.

2 Dale H (1946) Pilgrim Trust lecture to the National Academy of Sciences of the USA.

3 Rivett D (1947) Science and Responsibility. Canberra University College Commencement Lecture, 27 March.

4 Pettigrew P cited in Klein N (2001) They call us violent agitators. *The Guardian*. 23 March.

5 See Wackernagel M *et al.* (2009) Global Footprint Network, at <http://www.footprintnetwork.org/en/index.php/GFN/>

6 See de Soysa I and Gleditsch NP (1999) *To Cultivate Peace: Agriculture in a World of Conflict*. International Peace Research Institute, Oslo.

7 Kaplan R (2000) *The Coming Anarchy – Shattering the Dreams of the Post Cold War*. Random House, New York.

8 Shah A (2009) Today, over 25,000 children died around the world. Global Issues (web site). <http://www.globalissues.org/article/715/today-over-25000-children-died-around-the-world>

9 Roederer JG (1998) *Communicating with the Public, Politicians and the Media*. COSTED Occasional Paper No 1, July. UNESCO, Paris.

10 House of Lords Select Committee on Science and Technology (2000) *Science and Technology – Third Report*, UK Parliament, London, 23 February 2000. <http://www.parliament.the-stationery-office.co.uk/pa/ld199900/ldselect/ldsctech/38/3801.htm>

11 Eurobarometer 55.2 (2001) European Community Directorate-General for Press and Communication. *Europeans, Science and Technology*.

12 UNESCO World Conference on Science, Budapest (1999) Science for the Twenty-First Century. <http://www.unesco.org/science/wcs/index.htm>

13 <http://www.unesco.org/science/wcs/eng/declaration_e.htm>

14 UNESCO World Conference on Science, Budapest (1999).

15 Gristock J (2001) Science and society: towards a democratic science: a report of the British Council seminar held at Moonfleet Manor, Fleet, Dorset, 11–16 March 2001.

Chapter 2

1 See Myers DG (1999) *Weekly Standard* 4 (10 May 1999).

2 For those who nevertheless wish to refresh their memories of the basics of writing good English try:

(i) Blaxell G and Winch G (1995) *The English Language Users Guide*. Phoenix Education, Melbourne.

(ii) Zinsser W (2001) *On Writing Well*. Quill (HarperCollins), New York.

3 For those who wish to acquire the basics of good journalistic writing, the 'bible' is still Evans H (1972) *Newsman's English: A Guide to Writing Lively, Lucid and Effective Prose*. Heinemann, London.

4 For a witty account of the crimes of bureaucratese and their pitfalls, see Watson D (2003) *Death Sentence: The Decay of Public Language*. Knopf, Sydney.

Chapter 3

1 Macnamara JR (1982) *Public Relations Handbook*, pp. 28–29, Margaret Gee Media Group, Melbourne.

2 Littlewood G (1998) CSIRO seminar paper on public awareness and crisis management. CSIRO, Canberra.

Chapter 4

1 An example is the development of public and political awareness of the threat posed by emerging infectious diseases, and the subsequent global reinvestment in a research field that had previously suffered neglect.

2 Fisher N, Peacock A and Cribb JHJ (2007) Reading the public mind: a new approach to improving the adoption of new science and technology. *Australian Journal of Experimental Agriculture* **47**, 1262–1271.

Chapter 5

1 <http://www.timelesshemingway.com/faq/faq7.shtml>

2 A specialist news service delivering science stories to media in Australia, New Zealand and internationally is <www.scinews.com.au>

Chapter 6

1 Gray G (2000) *Engaging Politicians and the Community in a Dialogue for Science*. Federation of Australian Scientific and Technological Societies (FASTS), Canberra.

2 Market Attitude Research Services (MARS) (2001) Survey of Australian Federal MPs' Expectations for Science. MARS, Sydney.

3 Keir M (2001) *Pers. comm.*

4 Market Attitude Research Services (MARS) (2001) Survey of Australian Federal MPs' Expectations for Science. MARS, Sydney.

5 Fisher NI and Kordupleski RE (2000) *Focus Group Study of Scientific Information Requirements by Federal Politicians.* CSIRO, Canberra.

6 Australian Parliamentary Library (4 January 2001) *Senators and Members, by Levels of Qualification.* APL, Canberra.

7 Roederer JG (1998) *Communicating with the Public, Politicians and the Media.* COSTED, Occasional Paper No 1, July. UNESCO, Paris.

8 Gascoigne T (2001) *Report on Science Meets Parliament.* FASTS, Canberra.

9 Market Attitude Research Services (MARS) (2001) Survey of Australian Federal MPs' Expectations for Science. MARS, Sydney.

10 Fisher NI and Kordupleski RE (2000) *Focus Group Study of Scientific Information Requirements by Federal Politicians.* CSIRO, Canberra.

11 Gray G (2000) *Engaging Politicians and the Community in a Dialogue for Science.* Federation of Australian Scientific and Technological Societies (FASTS), Canberra.

12 Roederer JG (1998) *Communicating with the Public, Politicians and the Media.* COSTED, Occasional Paper No 1, July. UNESCO, Paris.

13 Parsons W (2001) Scientists and politicians: the need to communicate. *Public Understanding of Science* **10**, 303–314.

14 Hartomo S and Cribb JHJ (1999) *LIPI Marketing Communication Strategy.* Indonesian National Science Institute (LIPI), Jakarta.

15 Conclusions drawn by participants following a LIPI science briefing in the Indonesian Parliament on earthquake hazards.

16 de Soysa I and Gleditsch N (1999) *To Cultivate Peace – Agriculture in a World of Conflict.* Oslo Peace Research Institute, Oslo.

17 Scherr S and McNeely JA (2001) *Common Ground, Common Future: Using Eco-agriculture to Raise Food Production and Conserve Wild Biodiversity.* 8–10 November, IUCN – The World Conservation Union, Gland, Switzerland.

Chapter 7

1 House of Lords Select Committee on Science and Technology (2000) *Science and Technology – Third Report*, UK Parliament, London, 23 February 2000. <http://www.parliament.the-stationery-office.co.uk/pa/ld199900/ldselect/ldsctech/38/3801.htm>

2 The British Council (2001) *Science and Society: Towards a Democratic Science.* Oxford, UK.

3 Wynne B (2006) Public engagement as a means of restoring public trust in science – hitting the notes but missing the music. *Community Genetics* **9**, 211–220.

Chapter 8

1 A current list of internet users is available at: <http://en.wikipedia.org/wiki/List_of_countries_by_number_of_Internet_users>

2 The respective websites are <www.sciencealert.com.au> and <http://www.eurekalert.org/>

3 The world's media can be accessed through commercial media directories or sites such as <www.thepaperboy.com>

4 See <www.scinews.com.au>

Chapter 9

1 Gaskell G, Bauer MW, Durant J and Allum NC (1999) World apart? The reception of genetically modified foods in Europe and the US. *Science* **285**, 384–387.

2 House of Lords Select Committee on Science and Technology (2000) *Science and Technology – Third Report*, UK Parliament, London, 23 February 2000. <http://www.parliament.the-stationery-office.co.uk/pa/ld199900/ldselect/ldsctech/38/3801.htm>

3 Covello VT (1989). Principles and guidelines for improving risk communication. In *Translation of Risk Information for the Public.* (Ed. EB Arkin) pp. 127–135. Plenum Press, New York.

4 Adapted in part, with acknowledgement, from APEC (2001) *Communicating about Agricultural Biotechnology in APEC Economies: A Best Practice Guide.* APEC, Singapore.

Chapter 10

1 Cullen P and Markwort K (1996) Communicate, publish or perish – paradoxes in the emerging paradigms. Fourth International Conference on the Public Communication of Science and Technology, Melbourne, 11–13 November.

2 Gascoigne T and Metcalfe J (1997) Incentives and impediments to scientists communicating through the media. *Science Communication* **18**(3).

3 Searle S (2009) Australia's scientists are willing but not able. *Pers comm.* 24 June 2009.

4 Based on Cribb JHJ (2009) Media Course for Scientists. Julian Cribb & Associates. Also CSIRO (2001), Guidelines on Public Comment; CSIRO (2001) Communicators Handbook; and The Royal Society (2000) Scientists and the Media: Guidelines for Scientists Working with the Media. <www.royalsoc.ac.uk/policy/index.html>

Chapter 11

1 Searle S (2009) Australia's scientists are willing but not able. *Pers comm.* 24 June 2009.

2 Based on CSIRO (2001) Guidelines on Public Comment. Note: this was an earlier version, used for a number of years before management changed its policy towards communication.

3 It is important that permission to communicate be reflected in other policies that govern staff behaviour, not simply the public comment policy; otherwise its effect will be nullified.

4 Unless there is some clear and known reward for communicating, even the best of policies advocating it may prove ineffectual. These need to be built into terms of employment, job descriptions, annual staff assessments and promotion criteria.

Chapter 12

1 Littlewood G (1998) CSIRO seminar paper on public awareness and crisis management. CSIRO, Canberra.

INDEX